LAND ROVERS IN MILITARY SERVICE

BY BOB MORRISON

ISBN 185520 2050

Published by

LRO PUBLICATIONS LTD.

Distributed by

Brooklands Books Ltd., PO Box 146, Cobham, Surrey KT11 1LG, England

INTRODUCTION

One sunny afternoon, with camera in hand, I walked into Aldershot's Buller Barracks to photograph a military Land Rover for a society journal. That fateful day was to be a turning point in my life.

The Sergeant Major smiled. Our conversation went something like this. *RSM.* "Well laddie, what sort of Land Rover do you want to photograph? Hard top, soft top, portee, long wheelbase, short wheelbase, 88, 101, 109, ambulance, lightweight, gun tractor - the choice is yours. *Photographer.* "Don't know - I thought a Land Rover was just a Land Rover!" How wrong can you be. That afternoon I started my photographic research of the military Land Rover in all its' guises, and I'm still at it more than a decade later.

In 1987, I approached the publisher of the fledgling Land Rover Owner Magazine with the idea of doing a feature on the military Land Rover. On reflection, John Cornwall decided that a short series on this specialist topic might be more feasible due to the variety of the different models then in service. The first Military Scene column appeared in the January 1988 issue, and it is still running today, more than five years later.

For more than sixty issues now, it has been possible to find a different subject or slant for my column each month. Over the last couple of years, many issues of LRO have covered two military orientated topics, and some months when space permitted, there have even been three. At time of writing I still have not covered everything I want to and have a wish list longer than this page. Like the Land Rover, the subject just seems to run and run and run.

This book is a compilation of some three dozen or so of these monthly LRO features. Should you the reader be unfamiliar with the military Land Rover story, these articles will give you a varied insight into the world's most versatile combat and support light utility vehicle. They also illustrate many of the roles of today's British Armed Forces.

The first Land Rover entered British military service forty-five years ago, and the marque looks certain to serve on for at least the next fifteen years. No other army vehicle in this class has anything like it's track record. Hopefully this compilation will adequately illustrate why the marketing men picked Defender as a brand name to take the humble Land Rover into the twenty-first century

Bob Morrison

COVER ILLUSTRATIONS

Front- GLOSTERS hard top Land Rover 110 FFR rigged for airmobile operations.
Back TL - Military Police Defender 110.
Back TR - 24 Airmobile Brigade 90 FFR
Back ML - Series III 109 Carawagon Command Vehicle.
Back MR - Parachute Regiment 109 on arrival at the Drop Zone
Back BL - Series III lightweight and 109 models side by side.
Back BR - Sixteen year old Forward Control 101 airportable ambulance in daily service.

COPYRIGHT © 1993 BY BOB MORRISON, LRO PUBLICATIONS LTD.
AND BROOKLANDS BOOKS LTD.

DISTRIBUTED BY

Brooklands Books Ltd., PO Box 146, Cobham, Surrey KT11 1LG, England
Phone: 0932 865051 Fax: 0932 868803

Brooklands Books Ltd, 1/81 Darley St., PO Box 199, Mona Vale, NSW 2103, Australia
Phone: 2 997 8428 Fax: 2 452 4679

Motorbooks International, Osceola, Wisconsin 54020, USA
Phone 715 294 3345 & 800 826 6600 Fax: 715 294 4448

CarTech, 11481 Kost Dam Road, North Branch, MN 55056, USA
Phone 800 551 4754 & 612 583 3471 Fax: 612 583 2023

CONTENTS

On Patrol in Ireland	4
Belgian Minervas	8
The JRA One Ten	10
A Bigger Splash	12
Colour Supplement Part 1	17
Red Arrows Support	21
An Unusual Sight	22
Solihull Awakes	24
Count the Flags, Not the Barrels	26
It's Good News Week	31
Four by Four Fighting Force	32
Desert Rats Head for Gulf	36
Rovers in the Gulf	40
Sultan's Forces	46
PODs on V8's for Kuwait	48
Operation Haven	50
Red on, Green on, Load gone - Air Drop	53
Iberian Connections	54
Falling From the Sky	58
Royal Yeomanry	60
Moving the Goalposts	63
Lightweights and One Tens	64
Ace Mobile	66
Colour Supplement Part 2	69
Danish Military Land Rovers	73
The Perfect Choice	74
An Original	76
Well Proven	78
An Early Retirement	80
Airmobile Gunners	84
Tough Guys	86
Marine Landing	88
While Bullets Fly	90
Airmobile Signals	92
RAF Regiment	94
Military Vehicles on the Market	96

Military Scene

On in Ir

◀ *Series III Soft Top of 1 Infantry Battalion.*

▼ *Series III Soft Top fitted for radio of 3 Infantry Battalion.*

▼ *Soft Top FFR of 6 Field Signal Company.*

▼ *One Ten Soft Top of 6 Infantry Battalion. Although aerial mounts are not fitted, this vehicle is actually an FFR.*

EIRE, the Republic of Ireland, in common with around 150 other nations, maintains a large fleet of military Land Rovers. The range of vehicles used is as varied as the duties of the Irish soldier, sailor and airman, with both Series III 109" and One-Tens currently in service.

Technically a neutral country, Eire has never been at war, but this does not mean that its defence forces are in any way lacking. The country is to the fore in United Nations peace-keeping forces around the world and has suffered losses out of all proportion to the size of its army. Irish soldiers, like Scandinavians and South Sea Islanders, are just as likely to be sitting at the wheel of an all-white Land Rover with a blue UN on the doors, whether the hot spot is the Lebanon or a tiny African republic.

Being neutral, Eire is anxious not to be tied to any one country or power block for its military hardware, so it is not unusual to find Toyota Land Cruisers alongside Land Rovers in the same batallion, but at present the Land Rover is still in the majority.

The Irish military Land Rover is not the same as its British cousin. Many of the Permanent Defence Force vehicles are purchased via dealers who obtain standard specification Rovers and upgrade them to the level desired by the PDF. For example, unlike British Army vehicles which on the whole are tasked to the defence of Europe, Irish Land Rovers are used mainly for on-road work in the Republic and the tyres must reflect this role. It is expected that for the bulk of its service in the field, the British vehicle will be committed to towing a loaded trailer or even an artillery piece off-road and must therefore have a strengthened rear cross

Bob Morrison

▲ *Irish Navy One-Ten Soft Top General Service.*

member. The Irish vehicle, whilst still likely to be used for towing trailers like any Rover, is unlikely to be tasked in the same manner and can cope with the standard cross members.

A typical Series III General Service Rover is a soft-top 109 with seating for three in the front and bench seats for eight more in the rear. A military pattern front bumper is fitted but the civilian style rear grab handles are usually retained. Having said that, some vehicles can be seen with military rear bumpers fitted to the civilian rear cross member with three bolts.

A NATO style towing pintle is fitted along with a standard trailer socket. The basic colour of Army Series III Rovers is a semi-gloss dark green, which is a touch greener than NATO green. The chassis is black and the bumpers and trim are galvanised. Number plates, which are of the same configuration as civilian plates, are fitted to the top of the front bumper and to the right rear body panel above the grab handle.

Tactical signs, painted on bridging plates are normally carried on the radiator grille and the left rear body panel. They consist of a shield bi-sected top right to bottom left, colour coded to depict command and unit, with the exception of corps vehicles which are single colour. Tilts are full length canvas of khaki or olive green colour.

Fitted for radio

Both soft and hard-top Series III Fitted For Radio (FFR) vehicles are used. Externally the only thing which differentiates between GS and FFR soft-tops is the aerial mounting fitted behind the drivers doors. Internally, the normal 24 volt battery box and radio table is fitted, as is the

▲ *Exterior of Series III FFR of 6 Field Signal Company.*

▼ *Hard Top Series III FFR of 6 Field Signal Company, interior.*

▼ *Typical One Ten GS Soft Top of Supply and Transport Corps Depot.*

Military Scene

cable tray type of radio mount fitted to British FFR Land Rovers. The Irish aerial mount is quite distinctly different from the British aerial mount occasionally found fitted in this position. Series III PDF FFRs do not appear to have any provision for the wing mounted aerial mounts normally found on NATO military FFR Land Rovers.

The hard top FFRs are of the window van type based on 109" chassis with military pattern rear cross members. Internally they carry the same radio and battery fit as the soft-top, and likewise have an aerial mount behind the driver's door. They have military rear bumpers, towing pintle, etc, and the number plate is fitted to the left rear panel with the Tac-sign fitted to the rear door.

Navy One Tens

The One-Ten is now in widespread service with both the army and the navy. Navy GS versions are navy blue with dark grey tilts, and have black chassis and wheel-arch cheeks.

The vehicle I photographed had a Series III type military pattern front bumper, military pattern rear bumpers and a rear towing pintle. The brake and indicator lamp lenses were protected by circular mesh guards and the front lights by a rectangular mesh guard. On this vehicle the Navy Tac-sign of a white anchor on a light blue shield was carried in the rear position only. Both high intensity fog lamp and reversing lamp are fitted to the Navy version. Rear compartment seating consists of two double bench seats each side.

The bulk of Irish Army One-Tens are soft-tops, although some hard-tops are used for specific tasks. Both GS and FFR versions are in service, the latter being distinguishable by aerial mounts on both wings and a short post mount behind the driver's door. Standard radio table, battery box and radio mounting rack are fitted to the radio version.

Two distinct front bumper types are to be seen, one being similar to the standard Series III pattern and the other being a normal One-Ten bumper with towing pin but with the addition of upper sections as per the Series III. From observation the second type seem to be found mainly on FFR Land-Rovers. Both GS and FFR types have military rear bumpers, towing pintle and mesh light guards as per the Navy version, but not all have fog and reversing lamps.

Tac-signs follow the same format as Series III, but some vehicles carry the front plate on the right side of the bumper and some carry it on the radiator grille.

When the bumper with tow pin is fitted, the number plate is fitted to the top surface, otherwise it is bolted to the front face. Rear number plates on all One-Tens appear to be fitted on the left side.

On some vehicles spare wheel mounts are an integral feature of the bonnet. Many One-Tens are finished in the same dark green as the Series III vehicles, but some are painted olive green similar to NATO green, possibly the latest ones.

A typical One-Ten hard-top is the Light Aid Detachment vehicle of the Supply and Transport Corps. It is similar to the standard One-Ten GS apart from the van type top and side-hung rear door. The rear panels have windows and a rack is bolted to the door pillars to take a pair of orange beacons and a swivelling spotlight. Internally, the vehicle would carry all tools and equipment necessary for running repairs to other vehicles in the formation.

▲ Typical One Ten Soft Top of 6 Field Company in FFR role.

◀ Hard Top One Ten Light Aid Detachment vehicle.

▼ Both these vehicles are One Ten FFR Soft Tops of 6 Infantry Battalion, although the rear Land Rover has no wing aerial mounts fixed.

By Bob Morrison

On patrol with 6 Infantry Battalion, PDF.

THERE have been many instances of robbery, kidnap and armed aggression in the Republic over the last two decades. As a result whenever a person with terrorist connections appears in court it is necessary to provide an Army escort for the police accompanying him.

In border areas, the Republic mounts security patrols just as the Royal Ulster Constabulary and British Army do north of the border. Following one notable armed robbery which netted the INLA some £500,000, it has also been deemed necessary for the PDF to provide armed escorts to all cash transit vehicles. In almost all instances Land Rovers would be used as the troop transporters.

I had the privilege of watching a typical patrol from 6 Infantry Battalion practising dismount and deployment from their vehicles. Unlike many military drills, this one is practised in the knowledge that the next one could very well be for real.

Whilst patrolling a lane in the border area this unit could chance upon a team of assassins fleeing from Security Forces north of the border and come under sustained small arms fire. In such an incident only fast instinctive reaction honed by endless training would save their lives!

A typical patrol consists of two Land Rovers each containing at least a driver and three soldiers armed with SLRs. One vehicle at least would be FFR and the other probably a GS. The FFR would be in constant touch via radio net with HQ and other supporting units.

At a given signal or the first sign of trouble, the drivers would bring their Rovers to a rapid controlled halt and within seconds the patrol would have dismounted and taken up covering positions to a pre-arranged plan using whatever cover the terrain provided. To prevent the vehicles being commandeered, the drivers stay close to the vehicles with the engine running and the door open and the patrol, unless in hot pursuit of targets, would remain close enough to remount in seconds. To aid rapid dismount, the tailgate is normally removed from both vehicles and the two soldiers at the back of the Land Rover cover the ground they have just passed over with their Self Loading Rifles.

Thanks

In closing I'd like to thank the Irish Secretary of Defence for permission to visit Permanent Defence Force establishments and take photographs, Commandant Vincent Whelan of Curragh Command, Commandant Sennen Downs and Captain Padraig Storan of 6 Infantry Battalion and Commandant Declan Carberry of the Defence Forces Press Office. All went out of their way to extend typical Irish hospitality and friendship. **Bob Morrison.**

▲ Soldiers of 6 Infantry Brigade dismounting and dispersing from one Ten GS and FFR Landrover ▼

▼ The Tailgate is removed to aid rapid dismounting.

BOB MORRISON'S MILITARY SCENE

Belgian Minervas

*A Minerva in the workshop alongside an Army Series III.
If you can't afford a hi-lift jack, try an M113 tracked recovery vehicle.*

REGULAR readers will know from last month's column that I spent a fair bit of time on Exercise Iron Hammer trying to track down the unique and elusive Belgian Minerva Land Rovers. I was assured by both knowledgeable military correspondents and army personnel, that these small Land Rovers with distinctive wings had been spotted but no way could I find them. We did however spot a Belgian short wheelbase Series III Military Police Landy on the autobahn home, which confirmed that the Belgians had brought Rovers with them.

Never one to give up easily, I contacted the Belgian Department of Defence who provided not only three photos of Minervas, but also a photo of a Military Police Series III similar to the one we spotted.

None of the photos was taken during the recent exercise but this doesn't detract from their interest. I cannot profess to be an expert on these vehicles, but our postbag constantly carries appeals for information on this subject, so I'll pass on what little information I've gleaned so far.

The Belgian car manufacturer Minerva, based near Antwerpen were granted a licence in the early fifties, to build Land Rovers using Solihull engines, gear boxes, etc, on locally made chassis with body panels similar to the standard Series I. However, as they did not have the facility to form the compound curve on the front side panels, they used simpler sloping wings. A fancy radiator grille was fitted with louvres either side below the headlights, and indicators and sidelights were mounted on the wing fronts just above bumper level.

No tailgate was fitted to these vehicles, a 2/3rds height fixed panel being fitted instead, which allowed a jerrycan rack and spare wheel to be carried on the rear of the vehicle.

A small brake light was fitted on each rear panel in a different position from Solihull vehicles, and Jeep-style indicator/convoy lights were fitted to the rear cross member. Apart from these minor differences, the Minerva looked identical to the standard Series I 80" and was powered by the same 1997cc engine.

Production was from 1952 to 1956, with the Belgian armed forces and police taking vast quantities. It is said that the Belgian army

A pair of Belgian Recce Minervas armed with 0.30cal machine guns.

ockpiled so many of these vehicles, that even in the early eighties, mint condition examples were still being issued to some specialist units. The Minerva was also available on the civilian market, but these vehicles probably didn't have quite the same lighting and spare wheel arrangement. At least three Minervas exist in the UK in good condition.

During the closing stages of World War II, the Belgian SAS made use of Jeeps fitted with armoured windshields, machine gun mounts and spare fuel containers. It would appear that these SAS kits were fitted to Minervas in the fifties; one such example, without guns, can be seen on Page 30 of our August 1988 issue.

Should I come across any further information in my travels, I'll keep you informed through the occasional Loose Ends column. However it is well worth keeping a look-out at the ARC Nationals this year, as I had the pleasure of a quick trip in a Minerva at Trentham Gardens last year. Maybe 84734 will make an appearance again this year.

Bob Morrison.

Rear end detail on the privately owned ex-Belgian Army Minerva at ARC Nationals.

Belgian Para Commando Battery Minerva on exercise in the UK, 1986. (Phono: Graham Napper).

Bob Morriso

The One

▲ *Truck, carryall, lightweight, senior commander, FFR, winch, One Ten Land Rover. (photo: Graham Napper).*

Front and rear view of unusual soft top with internal lockers, rear facing seat, heightened tilt and double racks for jerrycans. (Photo: Chris Jones)

THE AUSTRALIAN Army uses the One-Ten in three basic configurations, namely soft-top, hard-top and station wagon. At a quick glance they look like standard Solihull built vehicles with a fancy camouflage and nudge bars, but close inspection reveals many differences.

Plastic modellers have a world-wide information network which would put the KGB and CIA to shame. Via a contact in another part of the Northern hemisphere, I've been able to aquire a set of Royal Australian Electrical and Mechanical Engineers' specifications for the One-Ten. Both the soft-top (Truck, Utility, Lightweight, FFR, Winch, MC2) and the station wagon, (Truck, Carryall, Lightweight, Senior Commander, FFR, Winch, MC2) are covered. The role of the former is to transport four personnel and radio equipment and the latter is to transport eight and radio equipment.

The major difference with the JRA military One-Ten is that a 3856cc Isuzu 4BDI diesel engine is fitted instead of the Solihull diesel, but the company also offers the standard 3.5 V8 for civil and other military users. Choice of the Isuzu engine stems from experimental work at the beginning of the decade to provide a high payload 6×6 with a diesel engine for military use.

Transmission is standard Land-Rover LT95A and normal Land-Rover transfer case and front axle are used with a GKN/Salisbury rear axle. Steering is Gemmer worm and roller and a Winch Industries Thomas T800M 3.5 tonne drum winch is fitted as standard (I know a few British sergeant-majors who'd give their back teeth for a winch on every Rover). 7.50×16 tyres are fitted as standard. with the spare being carried inside.

Externally the most noticeable differences are the nudge bar and winch fitted in place of the front bumper and the oversize rear bumpers, peculiar to Australian Rovers of all marks, which double as jerrycan racks. Fuel capacity is 65 litres, which won't get you far in the outback, so spare cans are a necessity, as is the

Military Scene

JRA Ten

One Ten soft top from JRA's demo fleet.
(Photo: Chris Jones)

A line-up of One Tens about to leave the JRA factory.
(Photo: Chris Jones).

Two Australian-built One Tens during Operation Caltrop Force in California.
(Photo: Mike Perring, Soldier magazine).

winch for extracting Rovers from some of the most diverse landscapes in the world. As with the Series III, the tail-gate is dispensed with in favour of a half-height rear panel and aerial mounts are fitted to both body sides. Pioneer tools are carried on the bonnet and two lockers, fore and aft of the rear wheels, are let into each body side. The front lockers carry battery trays and are found on both soft and hard-tops, but not on station wagons. On both hard-tops and station wagons, a tropical roof panel is fitted to help combat the fierce desert sun.

Although the front light arrangement is similar to British military Rovers, additional NATO style blackout lights are fitted above the headlights. The rear lighting arrangement is unique to JRA vehicles, consisting of bar clusters mounted beneath and protected by the rear bumper/jerrycan holders. A standard NATO pintle towing jaw and 12 pin socket is fitted in the standard position and a 24 volt electrical system is fitted to all Fitted For Radio vehicles.

The distinctive camouflage pattern sported by One-Tens consists of black 'clouds' over broad sand-pink stripes on a green base colour. The green is an olive shade, not dissimilar to the old NATO green shade still found on British Rovers and I can best describe the sand-pink as flowerpot-like. The tilt camouflage colours match the body but the pattern is a series of hexagons overlaid with random spots of contrasting colours. Bearing in mind the diversity of earth colouration and vegetation types found on the Australian continent, the elaborate and unusual colourscheme is probably very effective as a disruptive camouflage.

The JRA One-Ten is a good piece of kit which acquitted itself well on the recent American/Anglo/Australian/Canadian joint exercise in California, but by all accounts the company has produced an even better military workhorse in the shape of the 6×6. Next month we'll take a good look at this Land-Rover which is making everyone sit up and think.

▲ *6×6 makes for exceptionally spacious ambulance version.*

▼ *Revised rear chassis arrangement is clearly visible.*

▼*6×6 on the left, is not only stretched lengthways, but in width. Conventional One Ten on the right.*

A b

THE BIGGEST problem in writing about Land Rovers is that the story is ongoing and new information is constantly being un-earthed or is made available. Just as the last issue was going to print I had a letter from Ray Habgood, Land Rover Australia's Engineering Manager, clarifying a couple of points that I was unclear on and correcting a few mistakes. A midnight phone call provided even more information, and as we only had space for a couple of Perentie 6×6 photos in the last issue, we felt a fourth part to the Australian Saga was in order. I promise this will be the last on the Antipodean Rovers — for a little while anyway.

The decision to fit the Isuzu diesel engine to Australian Land Rovers was spurred by the fuel crisis of the late 1970s. JRA say they needed a diesel to match the performance of the 4 litre diesels used in Japanese Nissan and Toyota vehicles. The Isuzu 4BD1 was first fitted to locally assembled Stage 1 vehicles in place of the V8 petrol 3.5 litre engine. Australia is, of course, a prime target for all Japanese vehicle manufacturers as the country drives on the left like Japan which obviates the need to convert standard production line vehicles to left hand drive, and transport costs are much lower than to the UK which is Japan's other prime RHD market.

To provide the necessary 550km range required by the Australian Army, a single side mounted 65 litre fuel tank is fitted to the basic Australian Army 4×4 vehicles. The rear bumper on the military One-Ten is actually the rear chassis cross member which has been moved back to provide room for the spare wheel which is slung from a winch type hanger operated by the wheel brace. The wheel is located behind the rear axle in place of the normal fuel tank, and the rear end of the chassis has been significantly modified accordingly.

I assumed that all vehicles had the half height rear panel, but I'm told that all cargo vehicles except cargo FFR's have tail-gates fitted as standard. Hard top and station wagon variants have a full height side hinged door in place of a tail-gate.

All vehicles are fitted with a 12 volt vehicle electrical system; the FFR vehicles utilise an additional and separate 24 volt radio electrical system. Military lighting comprises front and rear L.E.D. type blackout lights, wing mounted reduced headlights and a rear convoy lamp which shines on the white painted rear axle diff cover. Lighting mode is controlled by a fascia mounted master switch.

The Australian Army has now adopted the space age Steyr AUG 5.56mm rifle an accordingly, most Rovers are fitted with butt boxes

Bob Morrison's Military Scene

igger splash?

and stock clips for stowage of two Steyrs in the cab.

The Thomas T8000M winch is not fitted to all vehicles, only 20% of the basic Cargo and Cargo FFR vehicles are so fitted. However all Survey and Senior Commander's variants carry winches as standard.

Variants

4×4 Cargo/Personnel. The four standard soft top One-Ten versions are C/P, C/P with winch, C/P FFR and C/P FFR with winch. Up to eight troops can be carried on inward facing seats in the rear compartment; the seats can be folded up for the cargo role.

4×4 Regional Force Surveillance Unit — this One-Ten has a slightly higher payload than the Cargo/Personnel vehicle. An auxiliary 50 litre fuel tank increases the range to 1000km. All RFSU vehicles are configured as FFR but not all are fitted with winch. A single rear facing seat is fitted in the cargo compartment and the tilt frame is heightened to give adequate headroom (see photos on page 20 of August 89 issue). A

▼ *Perfect water crossing technique by Aussie driver? (It makes a great picture).*

▲ Mobile worshop.

▼ This hard body 6×6 is configured internally as a mobile workshop.

Military Scene

A bigger splash?

heavy duty bull bar with side protection bars is fitted and a high lift jack is carried as standard. Additional spare wheels and jerry cans are fitted as standard and the vehicles carry an engine driven compressor for tyre inflation. Heavy duty bias belted Goodyear Custom Xtra Grip or Olympic Track Grip tyres on split rim wheels replace the standard radial ply tyres.

4×4 Survey Vehicle — this is a hard top, FFR with winch vehicle. A Command Post derivative is also available.

4×4 Senior Commander's vehicle — a Station Wagon FFR with winch. It has seating for eight personnel and carries radio equipment in the rear compartment.

4×4 Personnel Carrier — the standard Station Wagon, it has seats for up to nine soldiers including the driver.

6×6 Truck, Cargo — available either with or without winch, this soft top can carry three in the cab and up to twelve troops in the rear (for fuller description refer to September 89 issue).

6×6 Air Defence Vehicle — externally similar to the 6×6 Cargo, this vehicle is designed to tow either the British Aerospace Rapier or Bofors RBS70 SAM (surface to air missile) systems. Racks for reload missiles are fitted and the rear bed is recessed for easy loading of the system's tracker, camera and electronic units. The vehicles are FFR with winch and a tyre compressor is fitted.

6×6 Truck, Ambulance — the modular rear body is configured to carry either four stretcher cases, eight seated patients or 2 plus 4, as well as a medical attendant. A high capacity air conditioning unit is fitted in the module roof, linked to an engine mounted compressor. Various shelves, cupboards, lockers, etc., are built into the module and a large locker for the casualties' personal equipment is fitted over the cab roof. To provide communication between the attendant in the rear and the driver, a flexible sleeve connects the module to the sliding rear cab window.

6×6 Truck, General Repair — this modular body incorporates all necessary racks, bins and benches to provide a mobile workshop for two specialist tradesmen. Dependent on role, such as aircraft, vehicle or weapons repair, various tool kits can be supplied. Both body sides have large lift-up panels to provide shade and weather protection for external working and a large fold down work-bench is fitted to the left body side. 240 and 415 volt power is provided by an externally sited generator and 24 volt dc battery powered supplementary lighting is fitted.

6×6 Truck, Electronics Repair — similar to the General Repair vehicle, this modular body is fitted out for radio and electronics repair, but lift up side panels are not fitted as all work would be done inside the vehicle in this role.

6×6 Long Range Patrol Vehicle — this specialist vehicle has been designed for the Australian SAS and takes the Pink Panther/Desert Patrol Vehicle concept one stage further. Two prototypes are being evaluated at this stage and a firm order has been placed for a batch of vehicles which will incorporate any further special features identified during trials. The current configuration provides for driver and trooper up front and one trooper facing rear in the cargo compartment, with machine gun mounts provided for both passengers. 300 litre and 50 litre fuel tanks give an operating range of 1800 km, ten jerry cans can be carried and not less than four spare wheels are fitted as standard. Heavy duty bull bar and roll-cage are fitted for safety and both winch and high lift jack provide self recovery capability. A motor-cycle can be carried on a purpose designed frame at the rear of the vehicle. I hope to be able to give further details when the finalised design enters service.

Well, that just about sums up the Australian story to date, but somehow I don't think it will be long before I have to put pen to paper on this subject again.

Bob Morrison.

▼ *JRA 6×6 troop carrier.*

Land Rover Owner — the only magazine for Land Rover & Range Rover enthusiasts.

Every month at newsagents, £2.10

or, available by subscription from
LRO PUBLICATIONS LTD.,
THE HOLLIES, BOTESDALE, DISS,
NORFOLK IP22 1BZ
at £25.00 for 12 Issues

Take an annual subscription and we will send you FREE either: an L.R.O. T-Shirt or a Magazine binder.

- Twelve Issues of Land Rover Owner for just £25.00 - post free and avoiding possible price increases.
- Free membership of LROC (Land Rover Owner Club) - personal membership card means many extra discounts.
- Send a cheque today and get the next 12 issues of your favourite magazine delivered to your door.
- Membership of the International Off-Road Club.

Please send me the next twelve issues of Land Rover Owner, starting with the........................ issue and enrol me as an LROC member.

I enclose cheque/PO for £25.00 (UK) ★

Charge my Visa/Access

a/c no. ....

Expiry Date ....

Send me a free ....
(enter choice of gift)

Post your complete form to:
LRO Publications Ltd.,
The Hollies, Botesdale,
Diss, Norfolk IP22 1BZ

Name ....
Address ....
....
....
Signed ....
Date ....

★ Overseas rates available on request.

▲ 1a Hard top 110 and soft top lightweight 88.
▼ 1b Soft top 110 FFR of 210 SIGS lining up to drive into an RAF Chinook

▲ 2a Recce Rovers of the RAF Regiment heavily camouflaged with scrim and toting General Purpose Machine Guns.
▼ 2b Royal Marine winterised DEFENDER 90 and 110.

▲ 3a RAF Regiment 127 Rapier Tractors defending a UK airfield.
▼ 3b Early Series III four stretcher ambulance in it's twenty-first year of service.

▲ 4a Dutch soft top Series III 109 FFR draped with a camouflage net.
▼ 4b Norwegian hard top series III photographed in 1992.

By Bob Morrison

Red Arrows support

THIS MONTH, prompted by a transatlantic phone call from modelling colleague and LRO Club member Robin Craig, I'm concentrating on just one particular vehicle. Robin, who is a member of the Canadian branch of the International Plastic Modellers Society, specialises in building Land Rover models and asked for details of a specific RAF vehicle.

Each summer nine pilots in British Aerospace Hawk jet trainers, thrill millions of people with their unmatched team aerobatic skills at over a hundred displays and airshows around Britain and Europe. Despite fierce competition from around the world, the Royal Air Force Red Arrows are the elite, but without the support of their 26-man travelling ground crew, the pilots would not be able to carry out up to three performances a day at locations hundreds of miles apart.

Last August, I had the privilege of going airside at Exeter Airport to watch the ground crew preparing the Hawks prior to a display. In addition to being refuelled, each aircraft has to have its oxygen bottles re-filled, dye for producing coloured "smoke" added, and a myriad of other tasks completed before The Team climb into their cockpits. The trolleys, tool boxes, spare parts etc, required to keep the Hawks in the air travel from location to location in a Hercules transport aircraft. To move things around on the ground, a lightweight Land Rover painted in Red Arrows colours, runs back and forth along the flight line. As well as being used as a gopher, the lightweight is occasionally used as an aircraft tug for the 4 ton Hawks, and indeed at Exeter last year the Fleet Air Arm borrowed it to move over 8 tons of Sea Harrier around the apron!

Actually the "First Line", as the travelling ground crew are known, use a couple of different Land Rovers painted in team colours at their base RAF Scampton, but usualy only one performs in front of the airshow crowds. 05KD07, the lightweight that I photographed at Exeter, is a left hand drive hard top. Apart from an amber airfield beacon and a four height tow pin at the rear crossmember in place of the usual NATO towing jaws, it is a bog standard airportable 88″ Series III. It is painted red all over, including underside and interior, has a white roof, white bumpers, white rear door interior and a white cheat line down both sides. The legend ROYAL AIR FORCE, in black letters, is applied over the cheat lines, and the Red Arrows unit badge is carried on both doors. On the left door are three Hazchem stickers and there is white warning stencilling under the right door handle. Wing mirrors are black, tactical white on black registration plates are fitted, and tyres are Michelin XCLs.

Should you wish to model this vehicle in 1/35th, you will have to scratchbuild the body on a shortened Italeri 109 chassis — not as difficult as it sounds. However in 1/67th scale, Gordon Brown's Cromwell Models resin soft top lightweight can easily be converted. Parked alongside any of the 1/72nd scale Hawks on the market, only a purist would notice the difference in scales. Use Letraset for the stencilling, beg or borrow a couple of RAF squadron badges from an aircraft kit, and use squares of spares box decal to represent the Hazchem stickers — in 1/76th it is virtually impossible to pick out fine detail. Add a P. P. Aeroparts oxygen trolley and you've got the makings of a neat mini diorama.

Unfortunately there will be no Exeter Air this year, but I should be lucky enough to see the Arrows perform at at least half a dozen other venues this summer. Maybe team manager Andy Stewart will have procured a Ninety for this season — I know he's trying hard as he feels it would be just right for the job, not to mention the comfort of the ground crew who travel thousands of miles around the country and continent in the lightweight. If he's successful, I'll be there to record it — give me a shout if you see me.

Thanks to Brian Highley (introverted airshow organiser), Squadron Leader Andy Stewart (RED 10, 1989), "The First Line" and last but not least "The Team."

Bob Morrison

The Honorable Artillery Company

An unusual

ONE Saturday morning early in March, your fearless reporter woke to find clear blue sky, still air and a respectable totally unseasonal air temperature. Just right for a spot of photography.

Trouble was my programmed shoot had been cancelled and I was at a loose end. Not wanting to miss the first real good weather of the year, I left the girls tucked up in bed and set off for Salisbury Plain on the off chance that someone might be doing somthing unusual with a military Land Rover.

I knew in advance that several artillery units, including some old aquaintances, were booked in for live-fire exercises, so there should be something around worth photographing. However, when I checked the "pinks" with the range safety staff, I spotted something really unusual the oldest regiment in the British Army, were scheduled to fire 25 pounders. With a little bit of luck I might get a shot of a Land Rover moving one of their guns.

No doubt many of you are wondering what is so unusual about a Land Rover towing a 25 pounder. Well firstly the 25pdr was retired from front-line British Army Service in 1967 after an active 27 year life, in favour of the Pack Howitzer which itself has been replaced by the 105mm Light Gun. However, a few Territorial Army and Officer Training Units still use this fifty year old piece of ordance, which is still devastating in the right hands.

Secondly, the gun weighs 1800kg and is un-braked, which makes it a good bit heavier than the allowable civilian towing limit of 1500kg un-braked. Thirdly, it is considerably heavier than the quoted 500kg off-road towing capability of even the Ninety and One-Ten.

Lastly, the gun is longer, wider and heavier than a 109" Land Rover! However, the army knows the Land Rover's real limits, and uses it accordingly, so there was every possibility of finding Rovers replacing heavy Bedfords for moving the guns if the going was soft.

Having checked the given location for the HAC without luck, I was checking my grid references with the range staff when a small convoy of Bedfords led by a Series III 109 hard top trundled down the range road. I nearly dropped my camera in excitement when I spotted a Series III in the middle of the convoy towing one of the 25pdrs at reasonable speed.

Pausing only to take a couple of quick shots, in case they turned into the Danger Area and I lost them, I jumped into my car and followed them down one of the range roads — we were doing a steady 30mph.

After a couple of miles we turned off onto a rough track, without dropping speed noticeably, before the Bedfords and the tiny Land Rover towing its gun, broke off over rough ground.

Knowing their proposed grid reference, I shot off down the track to get ahead of them so that I could get a shot of this rare combination off-road.

On checking with the HAC Lieutenant in charge, I found out that one of the Bedfords had met with a mishap, and one of the Land Rovers had been pressed into service as an emergency gun tractor. However, it is usual for the HAC to move their guns around on the battlefield by Land Rover if the ground is soft, as the Rover is less prone to bogging than the Bedford. If, as occasionally happens, a vehicle does get bogged in really marshy ground, it is a lot easier to dig out two tonnes of Rover than five tonnes of Bedford.

BY BOB MORRISON

sight

- Series III 109 tows 1800kg of fire power cross country — the Land Rover's stated off road towing capacity is 500kg.
- Series III in use as a command post.

In addition to the Rover towing the gun, several other Land Rovers were in use that day. A couple of Series IIIs tucked away in nearby wood, one GS and the other FFR, were used as a Command Post complete with radio links and map tables set up in an awning attached to the rear of one vehicle. Another Series III nearby was used kitted out as a unit recovery vehicle, whilst another dished out lunch and copious quantities of Army tea.

Like many Territorial Army units, the Honorable Artillery Company is tasked to the British Army of the Rhine. The regiment's three operational squadrons would provide artillery support and act in a specialist recconnaissance role in Germany, whilst other companies act in the Home defence and Home Service Force Roles. However, to most Londoners the HAC, and in particular the Gun Troop, are best known for the ceremonial duties they undertake in the Capital.

Next month we'll look closer at this ceremonial side and the history of the Regiment.

Bob Morrison

MILITARY SCENE

◀ Road registered JRA 6×6 Cargo with winch.

Police Discovery ▶

Solihull awakes

HAVING had a bit of a go at Solihull in my February column, it was with some trepidation that I drove through the main gate at Lode Lane on Good Friday to face the management team. With no less than three directors, the Government and Military sales manager and the press officer lined up around the table, I was either for the high jump or an interesting day. Fortunately it was to be the latter.

In the past, to the outsider, Solihull has sometimes appeared as if caught in a time warp. Indeed it has been said that Land Rover were founder members of the Flat Earth Society but left as they felt it was getting too radical. However the winds of change certainly appear to be blowing through the corridors of Lode Lane. Maybe it's the success of Discovery, maybe the British Aerospace tie-in, or maybe a combination of both — I don't know for sure but you can almost taste the optimism in the air.

My two biggest moans about Land Rover were the apparent lack of positive publicity on the military side and the incredible delivery times quoted by dealers for the Discovery. I'm still not convinced that the message is getting through to military users, that Land Rover is still the "BEST 4×4×FAR", but the Government and Military Operations side feel that they are not only holding their ground but advancing.

At our meeting, it came out that a special demonstration of the Discovery was laid on for the Ministry of Defence even before its UK press launch. Not so much to try to sell it to the military, but to let them see what all the fuss was going to be about.

Land Rover were keen to stress to the military that the Ninety, One-Ten and One-Two-Seven ranges would be complemented by the Discovery, but no way was the humble Land Rover that we all know and love going to be any more than temporarily upstaged by the new baby. Take my word for it, contrary to some rumours being spread, the military, agricultural and contractor's Land Rover has a long life ahead of it yet.

Of course every vehicle must evolve to keep abreast of market requirements and Solihull has one or two ideas and tricks up its sleeve, but I'm not going to give the game away. However, a couple of interesting vehicles were spotted in the Special Vehicle Operations yard.

Most impressive was the Jaguar Rover Australia 6×6. If this beast had been around a few years ago, I doubt if the MoD would have plumped for the Reynolds Boughton in the 2 tonne category.

My request to take the JRA for a spin through the jungle was politely denied. understand that another JRA 6×6 variant will be on show at the British Army Equipment Exhibition — watch this space I also spotted a few of Solihull's conventional 6×6s plus One-Two-Seven crew cabs and flatbeds waiting for delivery to public service bodies.

Ambulances were also well represented in the yard with a batch of smart looking Quadtec bodied 127 military meat wagons ready for despatch to an African government client. Quadtec is a range of modular bodies tailored specifically for the One-Two-Seven range, which can be kitted out for just about any role by SVO.

A St. John's One-Ten station wagon with seating for six plus a stretched trolley was awaiting collection, and another pattern of military Quadtec ambulance was in the line-up. There was also a rather smart VIP station wagon with deluxe seats and luxury interior trim, complete with two seats in the back for bodyguards. Many role conversions which in the past would have been tackled by outside companies are now being handled in-house by SVO, which gives the customer the added guarantees of manufacturer's liability and service.

To get back to my other grouse — delivery times of Discovery. It is not often that a sales and marketing director admits

Solihull awakes

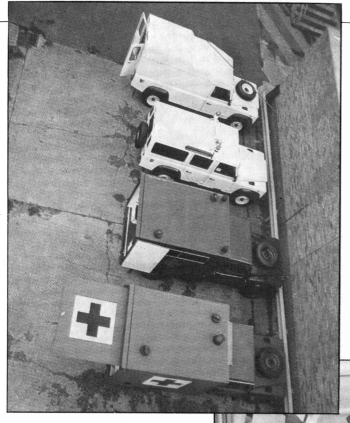

◀ Overhead shot of part of SVO yard with from top to bottom: new high roof One-Two-Seven, St. John's ambulance, military pattern ambulance for an African government, and Quadtec military ambulance.

Interior of a VIP One-Ten with seats for four plus two bodyguards. ▶

being caught out by trends, but John ussell is the first to admit that he and his am were caught well and truly on the hop y the colossal demand for Discovery. ithout boring you too much with statiscs, from government figures for the first ree months of 1990, three Discoveries ere sold for every two Shoguns or two for ery Trooper. In March the Discovery utsold the Shogun alone by more than o to one.

Compared with the first quarter of 1989, e number of vehicles sold in the 4×4 ctor at which the Discovery is aimed creased threefold. Discovery accounted r nearly half of these sales, and this is ith the Range Rover and Land Rover kept it of the equation!

With Discovery production now topping 0 a week, the waiting time is dropping, it demand is still increasing both in the K and on the continent. The French and lian markets are particularly buoyant. cidentally, Range Rover and 90/110 oduction and sales are not suffering, espite the unprecedented demand for the ew vehicle.

Although the Discovery has not yet been opted as a multi-purpose Staff Vehicle by e Ministry of Defence, Land Rover were le to confirm that the type is already in e with undisclosed units and the Queen's ght has some. A military contact has nfirmed my suspicions on which units e using Discovery.

The Police are also evaluating Discovery, not as a high speed pursuit vehicle, but in a totally new category. The Home Office has provided a standard equipment specification, which has to be an improvement on the hotchpotch of kit carried by Range Rovers of different Constabularies in the early days.

By the way, despite the Japanese manufacturers trying to gain a foothold in the Police market, they do not seem to be doing too well. As was pointed out, when you've succeeded in supplying all the vehicles in use in a particular class countrywide, heads are bound to turn when the competition gets the odd vehicle into service, but one swallow doesn't make a summer.

Having seen the Police-fit Discovery, and been hurled around the high speed test track in a Discovery by Don, the company chief demonstrator and walking encylopaedia on Land Rovers, I reckon the Police are onto a winner with Discovery. It looks good, too, in Police markings.

As well as showing me around SVO and the production lines, I had the chance to take a Discovery round the test track under Don's instruction, followed by a quick spin round in a 3.9 Rangey. Talk about being spoiled. Mind you, I did succeed in embarrassing myself by getting the Discovery stuck half way up the second flight of stairs — I blame the power steering, as it can't answer back.

As I think you'll have gathered by the tone of this article, I'm pretty optimistic about the future of Land Rover in all its marks. With nearly 8,000 vehicles in three totally different sectors of the 4×4 market rolling of Solihull's production lines each month, things are looking good.

Of course, there will always be a need for the military to test other vehicles to ensure that they are getting the best value for money for the taxpayer, but I reckon in the long run the competition will fall by the wayside.

Recently, one successful manufacturer of specialist multi-wheel military vehicles was asked if he could produce a lighter version of his vehicle with four wheels for a specific role. He told me that there was no point in attempting this as Land Rover already do it.

I'd like to express my thanks to George, John, Colin, Roger, Simon, Rod, Albert and Don for their help and hospitaltiy. Thanks also for the souvenir military specification pewter One-Ten model — despite pleas and threats from collectors, I will only exchange this for a full size Discovery.

Bob Morrison

BY BOB MORRISON

Series III 109's dwarfed by Foden wrecker

Count the flags, not the barrels

MILITARY SCENE

HERE WE go again! It's Spring Bank Holiday Monday and you lot are still enjoying yourselves at the ARC National while your humble scribe slaves over a hot word processor. Why do I always have to work Bank Holidays while everyone else puts their feet up? May Day was no better, as I spent a long weekend with ACE Mobile Force (Land) on exercise in Germany.

ACE Mobile Force (Land), or AMF(L) to use it's more common designation, is NATO's multi-national trouble shooting formation tasked with quick response to enemy threats on NATO's flanks. The primary role of the Force is deterrence, any potential enemy being aware that an attack on any one nation would constitute an attack on the full Alliance.

At present the countries contributing units to AMF(L) are Belgium, Canada, Germany, Great Britain, Italy, Luxembourg, Netherlands and the United States. Individually the national units are relatively small, but as a combined formation they present a formidable opponent. Hence the unofficial slogan — *Count the flags, not the barrels!*

The exercise for which I joined the Force was primarily an artillery joint exercise with all nations except Canada providing artillery batteries and/or mortar companies for two weeks of intensive helicopter deployment and live-fire training. The Canadian artillery element normally only deploys on Northern Flank exercises and Luxembourg, due to its small population, only contributes a reinforced infantry company, but this includes a mortar platoon.

As each national government funds its own equipment, the variety of vehicle used by the AMF(L) is second to none. Th British contingent uses every type of Lan Rover except the One-Two-Seven, and th Dutch field the One-Ten, but unfortunatel none of the other nations at present fiel Land Rovers.

The bulk of British Army Land Rove are Series III 109s for general duties an One Tonne 101 gun tractors, but the Nine and One-Ten are slowly coming o strength. One reason for the 109's so diering on so long is that they are fitte with winterisation kits. As AMF(L) regular excercies in northern Norway and nort eastern Turkey in winter temperatures times lower than −30 degrees C, the arcti heating kits are a necessity. Howeve many of the latest batch of Nineties an One-Tens ordered by the Ministry

...u can't beat the Ninety and One Ten'

Line up of Dutch Marines One Tens.
British One Tens and Italian Nuova Campagnolas.

...e will be winterised so it is a safe ...t by next winter many of the Series ... have been retired.

...biers in a quaint little German ...y, I was invited to join the lads on ...ar's winter deployment, but when ... where they are going and that *day-... emperatures* could be as low as ... I politely declined.

Honest opinion

...ways, I asked the troops in the field ...r honest opinion of the Ninety and ...n, and as usual all preferred it to ...ries III. Typical of most of the ...nts was this quote from a Captain ... Gibraltar Battery when asked why ... in a Ninety whilst the rest of his ...re suffering in 109s. "You can't beat ...ety and One-Ten — the sooner we all get them the better, so long as they are winterised" and "Yes, you can quote me on that."

Of course I agree with him, but I can't understand why the MoD is still buying naturally aspirated diesels when just about everyone else has gone over to Turbos. If the Germans and Italians trust their, mainly conscript, armies to look after Turbo diesels, can't we trust our professionals likewise?

The gunner of 5 Gibraltar Battery, 94 Locating Regiment RA field soft top One-Tonne 101s as gun tractors. Like Royal Marine 101s these vehicles have centre mounted winches, the cable feeding through the front cross member over a pulley to the left of the radiator grille. Due to their low weight but high power, they are ideally suited to airmobile artillery operations to remote areas, but they are getting a bit long in the tooth and will have to be replaced before long. Although AMF(L) does have some larger helicopters at its disposal, most of the day-to-day helilift capability is provided by the Pumas and Hueys of the Force Helicopter Unit. Even the Huey can lift the One-Tonne 101 when stripped for action, but the Puma is close to the limit with the Unimog and Hum-Vee. To lift the US 'Deuce and half', CH-53G Stallions or CH-47 Chinooks would have to be available.

The Royal Netherlands Marine Corps provide the Dutch contingent on this exercise. As time was short and I had already covered their One-Tens in detail (Feb 88 issue) I didn't spend long with the Cloggies, but I am happy to report that they are pleased with their One-Tens after two

*Above: British and Belgian military police with Land Rovers, a Bombardier and 4×2 Volkswagen 181.
Below: A Ninety in the field.*

years of hard use. Unlike the Dutch Army, the Marines bought their vehicles almost off-the-shelf from Solihull, and consequently have had little trouble with them despite the hard life that Commando Land Rovers tend to lead.

The other Commando contingent in AMF(L) is Belgian. The Belgian Para Commando Battery, part of the highly respected 1 Parachute Battalion descended from the Belgian SAS Squadron, until recently used the Belgian built Minerva variant of the Series One. After thirty plus years service, these hardy little Rovers have now been retired and the Canadian Bombardier has taken their place. The Bombardier is a licence-produced version of the Volkswagen Iltis with minor changes to meet Belgian specifications.

The dear old Minerva, like the Series One, was never the most comfortable of steeds over rough tracks, but the bum-numbing, teeth-shattering ride of the Bombardier/Iltis is little better, and with the tilt up it is just as cramped.

Although the Brits use the One Tonne to tow their artillery pieces, the 109 or One-Ten can be pressed into service if needed to do this job. No way can the Bombardier hope to tow a 105mm gun, so it is really only useful as a light utility and communications vehicle. Unimogs are used as gun tractors by the Belgians.

Hurt pride

The German troops tasked to AMF(L) are all crack Fallschirmjaeger (Paratroops). They also use the Iltis as a light utility vehicle, but the Mercedes G-wagen is appearing in small numbers as a command and communications vehicle. The German army has never really used the Land Rover as to do so would hurt the pride of the German auto industry, but they have never really been able to design a suitable competitor.

The larger Unimog is a superb gun tractor, but its price and performance are a bit over the top for the job required of it. The G-wagen is slowly picking up military sales, but it is more in the Range Rover league and is seldom used as other than a command and communications vehicle. However, one excellent little vehicle fielded by the German Paras is the Faun KRAKA, a 4×2 light load carrier capable of being air-dropped with the troops to provide them with a degree of mobility at minimum airload penalty.

The American contingent also comprise

Report and photos: Bob Morrison

Above: One tonne FC 101 towing 105mm Light Gun.
Right: RAF 101 used as Command Post.

Left: Mixed group of German Unimog, Italian Iveco and British Land Rover ambulances.

paratroopers, from the famous 82nd Airborne. Their light untility is the HUMMER or Hum-Vee which is unlike any other vehicle in its class, and one of my personal favourites.

The Luxembourg contingent also uses the Hum-Vee to transport its mortar teams, but for the command and communications role they have resorted to the G-wagen. In the past they used Land Rovers including One-Tonne 101s, but when they were looking for a replacement in the mid-eighties Solihull had nothing to offer at the larger end of the range.

The last national contingent present were the crack Alpini mountain troops of the Italian Army. Once again national pride dictates that they use their own vehicles rather than Land Rovers, but it is interesting to note how similar to the 88 and 101 their Fiats are.

As well as artillery and infantry elements, Britain also provides logistic support, military police, radio squadron, force reconnaissance, medical, air defence, air support centre and helicopter support elements. All of these units use Land Rovers of various types from the 101 ambulances of old friends 16 Field Ambulance, to the hard top FFRs of the Red Hand Gang who provide communications links between aircraft and ground forces and the RAF hard top 101 used as a command post for the Anglo/German Force Helicopter Unit.

Despite the mix of nationalities and languages, AFM(L) works incredibly well as a team, and the esprit-de-corps is evident everywhere. However, the logistics of keeping all the different vehicles on the road must give someone a hell of a headache. It would be nice to think that, for once, national pride could be set aside in favour of common sense.

With Britain's 101 fleet fast nearing the end of its projected service life and many other nation's vehicles either nearing the end of their life or proving less than ideal for the tasks required of a modern, quick response force, this could be the ideal time for Land Rover to sell the idea of commonality to all member nations of AMF(L). With the Ninety, One-Ten, One-Two-Seven and now the 6×6 all proven in military service, readily airportable and suitable for extreme climate operations, Land Rover can fill every category up to 2.5 tonne.

Eight nations fielding Land Rovers in one formation — seems pretty unlikely, but wouldn't it make a great photo.

Removing a 101 Land Rover and 105mm Light Gun from the drop zone after an especially hard landing.

It's good new

IT'S OFFICIAL. The Ministry of Defence has placed an order for no less than 500 Ninety and 1,190 One-Ten naturally-aspirated diesels for the British Army. At £22 million, this is the largest order placed by the MoD with Solihull for five years.

By the time you read this, the first chassis should be rolling down the line at Solihull and the entire order will be completed within 9 months. When you add this order to the 950 vehicles already ordered over the past 12 months, plus spares and parts orders, Solihull has done something like £46 million worth of business with the MoD recently. Things are looking good.

In past columns, I've told you of the One-Two-Seven Ambulance and Rapier Tractor contracts, but been unable to give numbers. It is now official that 200 Rapier Tractors and 100 ambulances have been ordered — both sizeable quantities for specialist role vehicles.

I've covered the ambulance in service in past issues and hope to be able to bring you in-service photos of the Rapier Tractor by the next issue.

Medical and Air Defence troops using the One-Two-Sevens all seem very pleased with them. Military contacts have also confirmed my suspicion that the One-Two-Seven Rapier Tractors that I saw at Solihull a couple of years back are operational in Turkey.

The other recent 650 vehicle order was for V8 One-Tens. Until recently the only V8 One-Tens I was aware of were used by SAS, but there is no way that they would need another 650! My Parachute Regiment sources tell me that they are in line for some and I've been picking up vibes which suggest that the airborne gunners of 7 Para Regt RHA might be trading in their ageing one tonne 101s for some.

I recently had the pleasure of watching 5 Airborne Brigade mounting a full 15 aircraft drop on Salisbury Plain. Unfortunately for one FC101 Gun Tractor, this was to be it's last drop, as the platform smashed into the ground just a little bit harder than anticipated.

During it's service life it has probably been dropped at least twenty times, so it gave of it's best, but it really is time that these hardy little vehicles were pensioned off. However, until something better comes along, the Paras will have to keep rebuilding their vehicles like 55KJ70. When I last saw this Rover on Stanford training area last year, it had just piled in. Now rebuilt from the chassis up and sporting a new late registration, it was back in action again, jumping out of an aircraft 800 feet up, then towing nearly 2 tonnes of Light Gun plus kit off the Drop Zone.

With it's high power-to-weight ratio, relatively small size, but practical layout, the FC 101 will always rate as one of the best Gun Tractors ever for Airborne action. I for one, will be sorry to see it go, warts and all, as it seems unlikely that sufficient numbers will ever be required again to make it viable for Solihull or restart the line.

Book Review

Richard de Roos has sent me draft copies, in Dutch and English, of his forthcoming book "The Land-Rover in Dutch Military Service" and I'm pleased to say that it is a fascinating read at a reasonable price for a specialist book.

Richard is a committed Land Rover enthusiast and former professional soldier in the Dutch Army. He has made good use of his obvious love for the Landy and his undoubted technical skills as a military mechanic to analyse Dutch military use (and misuse) of Solihull's products over a fifteen year period. I'll leave it to Richard to comment further:-

"In 1990 the Dutch Royal Army will

One Ten in service with the RAF. Now the MoD have placed a massive new order with Solihull.

s week

begin replacing all of the 4500 Land Rovers which it purchased in 1975. This book is about the period between 1975 and 1990 and covers not only the Land Rovers used by the Dutch Army, but also those in service with the Dutch Royal Marines, Airforce and Royal Military Police. The book will appear in the shops as soon as the Army has made a start at replacing it's fleet with an as yet unknown successor. Up to now very little has been published about the Land Rover in the Dutch Language and nothing about the Dutch Military Land Rover, which presents the most interesting news at the moment because of potential silence over aspects of controversy."

Once the book becomes generally available, we will bring you an in depth review. I found the draft absorbing and the photographic content superb. It also briefly covers the history of the marque and comments on the competition, but it's on the technical side that it really opened my eyes to military procurement and mismanagement.

I'm sure in time this book will rate as one of the classics not only among Land Rover enthusiasts, but also among military procurers as a lesson in how not to do things. Enough said at this stage. Dutch publication is scheduled for 21st June and English for February 1991.

De Land-Rover in Nederlandse Militaire Dienst ISBN 90-9003299-1. Price: 32 Guilders approx Publisher EUROBOEK, Haarlem, Holland.

The Land Rover in Dutch Military Service ISBN 90-9003378-5. Price: £10 approx. Publisher EUROBOEK, Haarlem, Holland.

Apologies

Most contributors to specialist magazines cannot make enough from writing on their chosen subject to make a living from it and I am no different. The daytime job which pays the mortgage and feeds the family must come first, which means that evenings, weekends and holidays are spent researching and photographing military Land Rovers and paperwork is fitted in around more important things like taking the photos, writing the column and cataloguing slides.

Especially during the summer months, replying to letters tends to take a low priority, not through choice but necessity. To those of you out there waiting for a reply, please accept my apologies — it's 6.10am and I'm working on it next.

The British climate doesn't do much to help either. Everyone organises there events to take place between 1st May and 30th September, so there are always clashes and double bookings. Usually the events I really want to cover are at opposite ends of the country, so something has to give.

Two recent opportunities that I would have liked to have covered were an exercise with 24 Airmobile Brigade and Driving Force 90. Both were within days of each other, both clashed with other exercises or events and both were in the "must go" category. Can anyone out there advise on how to extend the 24 hour day.

Lastly, as you read this, I should be swanning around France in a V8 Discovery. I'm visiting the French Ministry of Defence's Army Exhibition at Satory, to size up the challenge for 1992. Discoveries are selling like hot cakes on the other side of the Channel, so it will be interesting to hear what French manufacturers think of it.

Hopefully my report will be in the August issue — that is if I bring the Discovery back, of course, I might just stay over there if the weather is good and the wine is flowing.

Bob Morrison

MILITARY SCENE

British Army Equipment Exhibition 1990

Four by four f

AS USUAL, Land Rover was well represented at the British Army Equipment Exhibition. Indeed the first sight to greet visitors as they left the main hall for the outside stands was the Land Rover display.

A prettily painted Ninety in unusual pink, earth and sand three-colour camouflage and kitted out with no less than three machine guns, fronted the stand. However, the vehicle which attracted most attention was the Australian pattern 6×6 SAS variant tucked away amongst the potted palms behind the Ninety. Regular readers of this column know all about the 6×6, but to many visitors this version came as a bit of a surprise.

With it's 2.5 tonne payload and seating for fourteen soldiers plus three in the cab, the One-Ten 6×6 is ideal as a cargo or troop carrier. It gave it's most impressive performance during the mobility display where it was used as a gun tractor for the Royal Ordnance 105mm Light Gun.

However, although the 6×6 stole the show, the whole range of Land Rover vehicles was represented and all put in superb performances. After the Exhibition, courtesy of Land Rover, I was able to get some hands-on experience on most of the demonstration vehicles. Some vehicles were standard off-the-shelf models with only minor specification changes to suit the military user, but the majority were configured to undertake specific military roles.

Ninety

The Ninety Patrol Vehicle, resplendent in three-colour camo, was a LHD V-8 FFR model with winch, kitted out with a single GPMG for the vehicle commander and pedestal mounted double GPMGs for the rear trooper. With bumper mounted winch, blackout lighting, jerrycan stowage and radio racks, this is a budget version of the successful Desert Patrol Vehicle. If you need to mount short range, high profile, deterrent border patrols, this is the vehicle for you.

The Ninety Gunship, mounting an M40A1 106mm recoilless rifle in the rear compartment, is a low-cost tank and armoured personnel carrier hunter. The combination of anti-tank gun on Series III lightweight Land Rover has been well proven in battle, particularly during the Gulf War. Now produced on the Ninety chassis, with modified rear body and distinctive split windscreen the Gunship gives even the poorest nation a cheap, reliable and uncomplicated self-propelled weapon. Six reload rounds are carried on the vehicle and if required radios, machine gun

BY BOB MORRISON

SAS trio.

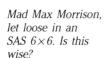
Line up of fighting Rovers for the nineties.

Mad Max Morrison, let loose in an SAS 6×6. Is this wise?

ghting force

mounts and smoke dischargers can be fitted.

One-Ten

Standard Land Rovers featured on many other manufacturer's stands as base vehicles for all types of communications and information gathering equipment. The most impressive of the Comms vehicles being the One-Ten FFR hard-top fitted with the latest British Aerospace Systems and Equipment FACTA. This Frequency Agile Compact Tactical Antenna (nicknamed the Magic Mushroom), operating over the full tactical VHF Band, is both compact and robust. It's low profile facilitates vehicle camouflaging and it is less prone to damage in the field than the conventional whip aerial. This One-Ten was finished in the three-colour brown, green and black colour scheme being introduced as NATO standard and as also used by France.

The General Service and Fitted For Radio versions of the One-Ten may be the workhorses of the military world, but it is the dashing Desert Patrol Vehicles, conjuring images of Stirling's Desert Raiders, that really depict the fighting spirit of the Land Rover. Of course, Colonel David Stirling actually used Jeeps and Chevrolet trucks, as he and his men wreaked havoc behind the German lines in World War II, but that was only because the Wilks brothers were about five years too late in creating the Land Rover!

Land Rover had two One-Ten Desert Patrol Vehicles on display — a Glover Webb bodied, left hand drive with winch and a genuine SAS right hand drive vehicle. The SAS vehicle had the standard weapons fit of three General Purpose Machine Guns, but the Glover Webb was equipped with a single forward firing GPMG and an Astra GECAL 12.5mm gatling gun on the rear pedestal.

Another variant on the One-Ten chassis was the POD ambulance. This system allows any non-dedicated GS vehicle to be quickly converted to a specialist role. Of specific interest to quick response units such as 16 Field Ambulance which is tasked to ACE Mobile Force, such vehicles provide a relatively cheap alternative to purpose built ambulances which spend much of their service life in mothballs.

One-Two-Seven

Although still very much a youngster as Land Rovers go, the 127 is fast carving a niche for itself in specialist corners of the military market. The integral bodied

MILITARY SCENE

BAEE 90

▲ *Ninety gunship.*

One-Two-Seven ◀ *Crewcab.*

ambulance version, conceived primarily as a rapid intervention airfield rescue vehicle capable of keeping up with the TACR2 tender, is now in widespread British service, including the Royal Navy and Royal Air Force. Other box bodied ambulances of different styles, on the same chassis, tailored to the customer's precise specifications are in foreign service. Having both driven off-road and been carried at speed around an airfield in the back of a 127 ambulance, I can vouch for their excellence.

The One-Two-Seven Crew Cab, with seating for six and a substantial cargo bed, was in action during the mobility display towing the Blindfire Radar for the British Aerospace Rapier surface-to-air missile (SAM) system. The Rapier Fire Unit was towed behind it's dedicated 127 SAM Tractor, which seats three and carries four reload rounds in addition to the four rounds carried on the Fire Unit. This 127 Rapier Tractor has even been purchased in quantity by the United States. Either the Crew Cab or Tractor version of the 127 can be used to tow the Royal Ordnance 105mm Light Gun or equivalent and any of the current generation of towed or pedestal mounted anti-aircraft systems is perfectly at home with the Land Rover — irrespective of nationality.

6×6

Well, what can I say about Land Rover's new military baby? It's everything I expected it to be and more!

There were no less than three of the beasts on show, two Troop/Cargo Carriers and a Long Range Patrol, complete with rear mounted motor-cycle. Both diesel and V-8 engines were fitted (I preferred the diesel) and all were equipped with winches. The cargo version made short work of towing the Light Gun at the Mobility Display, but no way did the course or conditions come close to taxing it to it's limits. I would like to see how it fares with really hard going when fully laden; though I'm sure it would fare just as well as a One-Ten.

I was able to drive all three off-road and on rough Long Valley tracks, though not to any great extent and felt perfectly at home with them in minutes. With rear seating for fourteen, the Troop Carrier is some size, but it still handles like a Land Rover. The "Mad Max" is such fun to drive that it's easy to forget that you're driving a small truck, but it is quite forgiving. With twin front facing GPMGs and that bloody great Royal Ordnance 30mm ASP canon behind you, you really feel invincible. For road use select 6×4 drive and high ratio.

Report and photos: Bob Morrison

▲ Pair of 6×6 troop carriers.

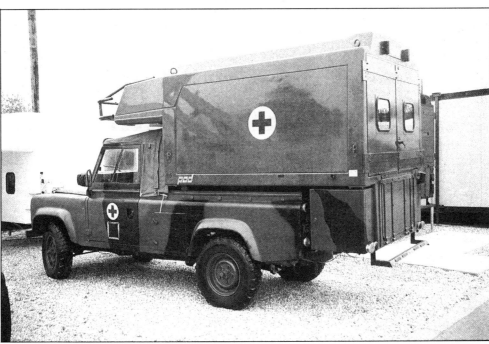

Demountable POD on the back of a 110 pickup. ▶

Off road, it's 6-wheel drive with 4F 1R through the two-speed transfer to suit conditions.

Wonder how soon it will be before Hereford aquire some.

Armoured

Three basic types of armoured Land Rover were on show. The Glover Webb Armoured Patrol Vehicle put on a low-key display at mobility. Having seen this type in the field in Northern Ireland, I know how it performs and the display just did not do it justice. Having recently seen how aggressively the French market their equivalent, this is one area where Land Rover could pull thier socks up.

In contrast, Penman's gaudily painted Hotspur Hussar six-wheeler put on a good display at mobility and the Internal Security variant on their stand looked quite the part. Despite little publicity, this version has quite a good overseas sales record. With seating for fourteen (including driver), gun ports, smoke dischargers and a machine gun turret, the V-8 6×6 Hussar is a cheap and economical light APC.

Shorts displayed two of their armoured vehicles on their stand. The S52 being the Patrol version and the S55 the APC variant of their One-Ten based family. Both vehicles are rated very highly, so it was a pity that their true potential was not displayed at Mobility.

The British Army FFLAV (Future Family of Light Armoured Vehicles) programme has identified a need for a vehicle in the armoured Land Rover class and even if disarmament continues at the pace some commentators predict, there will still be a need for security and patrol vehicles of this type. The Ministry of Defence have actually called this category of vehicle the "Rover Patrol". Foreign manufacturers are eager to have their vehicles licence produced in the UK to fill this category. It would be embarrassing if a continental rival pulled the rug out from under Land Rover.

Discovery

As I predicted, there was also a Discovery on show, albeit in a paramilitary guise. In other words it was army green with a police light bar and sirens. I'm sure that the Military Police, RAF Police and Ministry of Defence Police would all like to get their hands on this vehicle (who wouldn't), but I don't suppose they will get any until it has proved it's usefulness with civilian police forces.

Give it time.

22 Field Hospital Group on the move

Forward Control ambulance on the autobahn to Bremerhaven.
Line of FC 101 ambulances waiting for despatch.

BY BOB MORRISON

Desert Rats' One Ten FFR on Bremerhaven Docks.

Desert Rats head for Gulf

SINCE WRITING last month's Military Scene, the British Army commitment to the Multinational Force in the Gulf has increased dramatically. Our public information friends at 1 BR Corps and 1st Armoured Division in Germany have gone out of their way to provide facilities, for which I am most grateful. LRO club members and Land Rover enthusiasts in the forces have also done their bit.

After a contingent from 30 Signals Regiment, the next sizeable outfit to depart for the Gulf was the medical staff of 22 Field Hospital Group RAMC, who started moving in mid-September. As luck would have it, our paths crossed on an airfield in the early hours of the morning of their departure from the UK, and their Sergeants arranged for me to get photos of a couple of their One-Tens. Both were soft-top diesels, freshly brush painted "Sand". One was a GS, but the other was kitted out in the FFR role with internal radio racks and an aerial mount post on the right side of the body.

Modellers might find the following notes of interest. Large red crosses on a square white background were painted to the rear of both sides of Sand painted tilts with smaller crosses (of differing sizes) on circular backgrounds painted on the doors and right tail panel. The FFR also sported a callsign on both doors of yellow on black, black side locker doors with chalk mark stencils, a square Red Cross insignia on the radiator grille and a Union Flag sticker on the right rear bumper. The GS had an additional small circular Red Cross sticker on the front bumper. The rack lashed on top of the FFR is actually two tent or penthouse frames.

As most of you will now be aware, the bulk of the Land Forces element of the British response is being supplied by the 7th Armoured Brigade, who achieved immortality in World War II as Monty's Desert Rats. With two regiments of Challenger main battle tanks and a mechanised infantry regiment equipped with Warrior per-

MILITARY SCENE

MILITARY SCENE

Photos: Bob Morrison

22 Field Ambulance Group were amongst the first Army troops to take One-Tens on Operation Granby

Royal Army Medical Corps FFR One-Tens

sonnel carriers, the Brigade is one of the most formidable fighting formations of the British Army of the Rhine. I was in Norway with the Marines when the announcement was made, but as soon as I got back to the UK I started making arrangements to get across to Germany to photograph their vehicles before they departed out East.

New One Tens

The tank regiments, the Royal Scots Dragoon Guards and the Queens Royal Irish Hussars use Land Rovers as gophers (gopher-this and gopher-that) and until mobilisation were equipped with old Series IIIs. However, just prior to departure, much of their old kit was swapped for One-Tens, which has really pleased their MT Sections.

As I wandered round the RSDG vehicle sheds, recently arrived One-Tens were having the finishing touches applied to their new sand paint schemes.

1 Battalion the Staffordshire Regiment, who provide the infantry element, have also been equipped with a large proportion of One-Tens, but although I visited the regiment, I was unable to photograph their vehicles due to operational restrictions at the time of my visit. Most of the armoured regiments' vehicles carry no markings other than Union Flag stickers, but many of the Scot's Land Rovers sport the St. Andrew's Flag painted high on the left wing just ahead of the door pillar.

As well as the armoured contingent, supporting elements from the other corps were much in evidence on Bremerhaven Docks as the first vehicles were shipped out. 1st Armoured Field Ambulance took One-Ten GS soft-tops and FFR hard-tops, a few Series III GS and FFR soft-tops and a large number of 101 Forward Control ambulances.

Most vehicles were painted Sand, but a number remained in their usual green and black scheme. All medical Rovers sported the mandatory Red Crosses, but the One-Tonners had theirs flipped over to give a plain Sand appearance.

Bearing in mind that these vehicles were being shipped out a mere two weeks after the announcement that the Brigade would be involved, and all vehicles had to be overhauled prior to departure for a totally different climate, it is amazing that so many were painted before leaving Germany. It is anticipated that something in the region of 3,500 mechanically prepared vehicles will be moved out of Germany in little over three weeks from the off.

MPs on the scrounge

Another contingent was of Military Police Land Rovers. They also sported their normal temperate camouflage and had a mix of high and low visibility MP markings. All of the ones I spotted were either One-Tens or Nineties, mainly hard-top FFRs, but they also had a few soft-top One-Tens.

As always, the MPs were quick to grab freebie LRO stickers. I was happily snapping away when I realised that one vehicle was sporting an "LRO leaves the

▲ Military Police 110 — note the LRO sticket in the windscreen.

others standing" sticker as opposed to the "Tread Lightly" and "once a month" stickers that I'd been handing out. Turned out to be one that I zapped in the field on exercise last year and the sticker had stood the test of time! It's a small world.

At the time of my visit to BAOR, it looked likely that the Army Air Corps would also be supporting 7th Armoured Brigade, so I travelled south to Detmold to photograph their helicopters in two-tone camouflage. Subsequent statements by the Minister of Defence seem to rule out their deployment, but a week is a long time in politics, as the old saying goes.

When I visited the Regiment scheduled to go, their aircraft were already painted and prepared for departure, but their vehicles had only been prepared and masked prior to painting. At this time they were intending to take mainly One-Tens, plus their unit's integral FC101 ambulances. As with the armoured regiments, the technique seemed to be to mask the lights, number plates and markings with masking tape, the tyre walls with insulation tape and the windows with newspaper, before spraying overall with Sand. Touch up is done with a 2 inch paintbrush.

As is only to be expected by military off-road vehicles in everyday use, many of the older Land Rovers being despatched to the Gulf had bumps and dents in their panels. The Land Rover's forgiving aluminium construction allows the panel to be beaten back roughly to shape, and a couple of thick coats of paint are brushed on to camouflage the damage.

With a steel panelled vehicle, the chances are that the offending panel would have to be replaced at worst or corrosion treated and carefully re-painted at best. Makes you wonder how some procurement bodies can justify their claim that pretty pressed steel vehicles have a longer field life than the Land Rover.

To bring this instalment of the Gulf saga to a close, just a quick word of thanks to the LRO members who have contact me with snippets of info. Believe me, every little bit of information comes in useful in building up an overall picture.

Probably the most interesting titbit to come out of the region so far is the report of American troops using an olive green Discovery with Qatar registration plates. Thanks KF.

To those readers in the Gulf theatre, please keep me posted on developments, but don't take chances. The military can naturally be extremely sensitive about security, especially in an active operation like this. Even when photographing run-of-the-mill military (and civil) Land Rovers, I always ask for permission BEFORE taking off the lens cap. Always respect a refusal of permission, no matter how stupid the refusal may seem, and you won't land yourself in the soup.

STOP PRESS: Latest from Saudi Arabia — now confirmed nine V8i, 5-door Discovery.

MILITARY SCENE

Gulf Update by Bob Morrison

Rovers in the Gulf

AS I WRITE, the cease-fire has been signed, Kuwait is free again and the British Army, Royal Navy and Royal Air Force are preparing to come home. Thankfully, it looks as if all of our friends made it through in one piece and should soon be sitting back in their local with a welcome pint of ale or a stein of bier.

However, let us not forget the brave few who gave their lives in the Battle for Kuwait — our thoughts are with their families and friends. Last July, as I stood amongst the graves on the Somme, I thought that with the collapse of the Iron Curtain such sacrifices were over.

I was wrong!

To get back on a lighter note, Ross Floyd passed on part of a letter that he'd received from Warrant Officer John Rochester of the Royal Scots Dragoon Guards. We gave out John's address in the February issue for anyone wishing to write to lads in the field. Like the true gentleman that all Scots Sergeant Majors are, he passed on the letter from the 21 year old lady to one of the "younger and more available" of his lads.

By coincidence, the RSDG One-Ten on page 60 of that issue formerly belonged to John in his previous post, and he has "fond memories of thrashing it round the training areas and autobahns in Germany". He says "it had lousy pick-up acceleration and a top speed of 105 km/hr, but in defence it went all over the shop cross-country and had to be driven aggressively to get anywhere, especially after tanks had been through the vicinity."

By the way, one story which didn't hit the national newspapers was the one about the Land Rover and the burning Kuwaiti Oilfields! It appears that as Saddam Hussein's troops had set so many well heads alight, Red Adair had his work cut out and suggested to the authorities that they should try to find additional help from the Yellow Pages.

To cut a long story short, the only people available were a bunch of lads from Bantry Bay, and they weren't too keen on the job as Kuwait is a dry country, but the Emir twisted their arm with a bounty of a million dollars per well successfully extinguished. To get them there, RAF Lyneham dispatched a Hercules to Shannon Airport and after drinking every pub dry on the road between Glengarriff and Limerick, the boys were eventually on their way out to the Gulf.

After a brief stop-over at RAF Akrotiri so that they could have a pee and top up with cans of Guinness at the NAAFI (Strike Command refused their request for a tanker aircraft full of stout to rendezvous over the Mediterranean as it was felt that keeping the Tornado and Buccaneer squadrons aloft was more important), Patrick and his lads were back on course for an airfield in north-eastern Saudi Arabia.

As the Herc touched down at Ras Mishab, the tail ramp dropped and a battered old Series II hard-top with 16 men and a ladder shot out of the back. They sped up to and across the border, just to the north of Khafji, heading for the nearest oil fire. Behind them, the Press corps struggled to keep up in a fleet of brand new Jap 4×4s with air conditioning and go-faster stripes.

When they got to the oil head, Patrick astounded everyone by heading straight for the hottest part of the conflagration. Before the Landy had even stopped, the fifteen hired hands jumped out, whipped off their donkey jackets and started to beat the fire out with them.

Meanwhile Patrick casually removed his safety helmet, capped the well with it and sat on top of it like a pixie on a toadstool, eating his sandwiches. With the fire safely out, Kate Adie and Martin Bell hustled in for an interview.

"You're a rich man now," said Kate to Paddy. "What will you do with your million dollars," said Martin? "Well after I've paid off the lads, I think I'd better see about fixing the brakes on the Land Rover," said Paddy as he washed down his sandwich with a slug of Moussey alcohol-free beer.

On that note, this seems to be a good time to ease off for a while on the Gulf coverage, although I promise to do a follow-up on the markings used in the Land Battle phase, when photographs become available. For those of you interested in more Gulf Land Rover coverage, we've put together a photo-book on the subject called BRITISH LAND ROVERS IN THE GULF.

Based mainly on my trip to Saudi Arabia just before the outbreak of war, it also uses photographs provided by Stan Standley and Laurie Manton from the Gulf and has a section on Rovers leaving from Germany and the UK. It is soft-bound, full-colour and available from good bookshops and the LRO bookshop. Cover price is £7.99 and the I.S.B. No. is 1-873564-02-3.

One parting shot. All civilian vehicles in use with the military were painted sand immediately prior to the outbreak of the Land War and the Coalition invasion marking of a black inverted chevron was added to the sides.

Andy Jack managed to photograph one

BOB MORRISON'S BOOK "BRITISH LAND ROVER

Hurriedly snapped by Andy Jack, this Discovery in Saudi Arabia has been over-painted in sand camouflage and has military number plates. Note, too, the inverted chevron, denoting a coalition force vehicle.

of the Discoverys painted in this fashion. Even the rear windows were over-painted, which makes it look like a van conversion. Also of note is the military registration plates fitted instead of the Saudi civil ones. Presumably the Army purchased the vehicles before the shooting started. I doubt if Collison Damage Waver premiums cover for landmines.

THE GULF" IS AVAILABLE NOW FROM LRO BOOKS

MILITARY SCENE

By Bob Morrison

Batus and the red top Rovers

A LARGE number of the regular readers of this column are military modellers. By their steady flow of letters and phone calls they provide me with inspiration, and more importantly information, on aspects of military Land Rovers that I cannot hope to cover myself.

As I think I've mentioned before, the international modelling grapevine is quicker and more efficient than any government intelligence service. This month a couple of friends in the Canadian branch of the International Plastic Modellers Society have put pen to paper and supplied photographs of British Army Rovers in Canada. I'll hand you over now to Gary Soucey and Robin Craig.

The British Army runs a large tank training facility in Suffield, Alberta, Canada. BATUS (British Army Training Area Suffield) was opened in 1971 and covers about 600 square miles. The base is run by a staff of 200 permanent UK personnel on two to three year postings and another 100 temporary UK staff on three to nine month postings. A civilian staff of about 200 Canadians provide support for the unit.

Even though BATUS is primarily a tank training unit it has a large permanent fleet of Land Rovers. In January 1990 Canada switched to unleaded fuel and as a result a decision was made to change over the Land Rover fleet to diesel. As a result nearly all of the original Series II and Series III 109s and the lightweights have gone, although occasional vehicles like 09KA76 and 16KA37 have survived. Also at least one Series III ambulance survives.

Apart from adventure training vehicles, BATUS Rovers can be divided into two groups; those used by visiting formations, and those used by umpires and the range safety staff. The two types are easily distinguished, as although all of them are painted in a sand and brownish green camouflage scheme, umpire and range safety vehicles have their upper bodies or canvas tilts painted a fluorescent orange red. All vehicles carry their radio call signs on the doors, on the front of the roof and at the rear. Call signs are painted in white on a matt black square. Much of the time, although stencils are used, the results look uneven. A three digit number, believed to be a fleet number, is painted on the sides at the rear.

The third group, used for adventure training away from the base, are painted a deep Royal Blue. They carry no military markings other than their registration plates. The blue Rovers, along with Bedford 4 tonners, are used on the annual ARCTIC ROLLER winter exercise when the staff go to the remote North West Territories.

We would like to express our thanks to Staff Sergeant Houghton and Major Thatcher for their assistance.
ROBIN CRAIG and GARY SOUCE

PHOTOS

Photographs A and B show a typical BATUS hard top FFR Ninety as used by visiting British regiments. The sand paint appears to be the same colour as that used in the Gulf, but the green has a more brownish tint than the standard NATO green, which is visible at the front of the bonnet and around the rear door hinges where the camouflage scheme has worn off.

Photographs C and D are of one of the few remaining Series III soft top GS 109s. It is parked between a pair of hard top One-Tens. It has a civilian pattern rear cross member and front bumper and unlike the One-Tens has a brush painted finish.

Photographs E, F and G are of range staff One-Tens with both hard and soft tops. All are FFR as is only to be expected of safety vehicles. The soft tops have home-made wooden rear doors, and their call signs are painted on the front tilt as well as on the doors. Note that although (40) and (3B) are sprayed, (77A) is brush painted. (3B) has a half length tilt to allow observers to stand in the back — note the grab rails.

Photograph H shows another of the remaining Series IIIs. However, unlike 16KA37 in photos C and D, this one has

△ *Typical Bacus FFR90*   Photo A △

▽ *Photo C*   Photo B ▷

One of the remaining Bacus Series III soft top GS 109s
▽

Photo D ▷

been sprayed. Once again a civilian rear cross member is evident.

Italeri's 1/35th scale Land Rover kit is ideal for either of the Series III 109s, the only minor modifications required being the addition of mudflaps and towing jaws, and the removal of part of the body side to represent the stowage compartment behind the passenger door on (57). The JB Models kit of a soft top 109 is ideal for the 1/76th scale modellers. Both models are available through the LRO Model Shop.

More photos over page ▷

MILITARY SCENE

MILITARY SCENE

◁ Photo E

▽ Photo F

Above and left: Bacus range staff One-Tens with both hard and soft tops.

△ Photo G

Below: Another surviving Series III

Photo H ▷

WHEN ALL ELSE FAILS CONSULT THE LAND ROVER FACTORY MANUAL!

BROOKLANDS TECHNICAL BOOKS

Brooklands Technical Books has been formed to supply owners, restorers and professional repairers with official factory literature.

The following are available on Land Rover and Range Rover models:

Land Rover Ser.1 workshop manual	4291
Land Rover Ser. 1 1948/53 parts catalogue	4051
Land Rover Ser. 1 1954/58 parts catalogue	4107
Land Rover Ser.1 instruction manual	4277
Land Rover Ser.1 & II diesel instruction manual	4343
Land Rover Ser.II & IIA workshop manual (part 1 engine)	AKM8159
Land Rover Ser.II & IIA workshop manual (part 2 chassis)	AKM8159
Land Rover 2/2A/3 1959-83 owners' workshop manual	
Land Rover 2A/2B forward control parts catalogue	608218
Land Rover 2A/2B instruction manual	LSM641M
Land Rover Ser.III workshop manual	AKM3648
Land Rover Ser.III owners' manual	607324B
Land Rover Military (Lightweight) Ser.III parts catalogue	
Land Rover Military (Lightweight) Ser.III user manual	608180B
Land Rover 101 1 tonne forward control workshop manual (soft & hard cover)	RTC9120B
Land Rover 101 1 tonne forward control parts catalogue	6082943
Land Rover 101 1 tonne forward control user manual	608239
Range Rover 1970-84 workshop manual (edn. 7)	AKM3630
Range Rover (2 door) owners' handbook	606917

From specialist booksellers or, in case of difficulty, direct from:

Brooklands Books Ltd., PO Box 146, Cobham, Surrey KT11 1LG, England.
Telephone: 0932 865051   Fax: 0932 868803

Brooklands Books Ltd., 1/81 Darley St., PO Box 199, Mona Vale, NSW 2103, Australia.
Telephone: 2 997 8428   Fax: 2 452 4679

LRO (Mail Order Books), Bridge Farm, Thwaite, Bungay, Suffolk NR35 2EE
Phone & fax: 0508 458123

MILITARY SCENE

By Bob Morrison

△ Omani GPMG gunner — his raised seat gives a greater field of fire.

◁ Omani 109 in Gulf War markings in Kuwait City.

▽ Sultan of Oman's Air Force Series III 109 Rapier Tractor.

△ *Sultan of Oman's forces One-Ten Desert Patrol vehicles.*

Sultan's forces

EXPORT SALES have always played a major part in the success of the military Land Rover but many traditional markets, particularly in Africa, have now all but disappeared. Much of this is due to nations having gained their independence (not only from Britain) and as a consequence their economy has become less stable.

In the seventies and eighties the market became flooded with mass produced, mainly Japanese, 4×2 and 4×4 pickups which many saw as a cheap alternative to purpose-designed off-road vehicles. In addition, corruption is rife in many developing nations (and the occasional developed nation), and Solihull refuses to get involved in the practice of paying "commissions".

However, one overseas market which still holds up is the Middle East. In the oil-producing Arabic nations, where money flows freely, and the values of honesty and trust are still held in high regard, the Land Rover marque is held in esteem.

Despite vast numbers of Japanese 4×4s being unloaded in the Gulf region, mainly to offset oil purchase balance of payments deficits, elite and specialist Arab military forces have a preference for the Land Rover.

The Sultanate of Oman, at the mouth of the Gulf, is typical of many nations in the Region in it's use of Land Rovers. By no means all of their off-road utility vehicles are produced at Solihull, but a large proportion are. Generally their Land Rovers tend to be better equipped than the average British Army vehicle and are often used in specialist roles.

With a population of only 1.5 million and a land area almost as great as mainland Britain, 95% of which is desert and wasteland, Oman's armed forces must have a high degree of mobility to patrol their borders. It is in this region that the DPV (Desert Patrol Vehicle) is used to its full potential on long range patrols under the scorching desert sun.

In the seventies the British Army, including the SAS, and the Royal Air Force assisted the Sultan's Forces in defending his country from communist rebels funded by neighbouring South Yemen and the friendships forged between the two armies last to this day. With only minor variations, the Omani One-Ten DPVs are identical to the current British SAS version.

The Sultan's Air Force use the British Aerospace Towed Rapier for air defence; the Land Rover being the normal towing vehicle. Their Series III soft-top FFRs are almost identical to British Army 109s, but due to the vast areas covered, enhanced communications are carried.

The Rapier Tractor in the accompanying photo clearly shows the towing and carrying capabilities of the Land Rover off-road. The fire unit weighs in at over 1.25 tonnes and the combined weight of the four missiles in their containers in the back of the vehicle is around 300kg. In addition, an optical tracker can be carried in the rear and a three man fire team travel in the cab. Two more Land Rovers are sufficient to carry the other six men of a Rapier detachment. One Rover can tow a radar set and the other carries personal kit and tows a trailer with further missile re-loads.

During the Gulf War, Omani troops were some of the first into Kuwait City. The Series III 109 soft-top with invasion Chevrons was photographed outside the Omani Embassy in Kuwait shortly after the liberation of the city. Note the sand tracks, jerrycan rack, aerial mounts and plethora of kit.

Photographs courtesy of British Aerospace.

MILITARY SCENE

△ *Factory fresh V8 110 Hi-Cap Defenders awaiting the next batch of POD ambulance bodies.*

PODs on V8's fo

MILITARY SCENE

WITH THE Gulf War cease-fire now formally signed, attention is turning to the rebuilding of Kuwait and the resettlement of the Kurds in Northern Iraq. Naturally, the Land Rover is playing a major part in both operations in military and civilian service. This month we will look at the POD ambulance and next month I hope to report on the Royal Marines' involvement on Operation Haven.

Examples of the POD have been successfully trialled by specialist units of the British Army. Regular readers of this column will have seen it in military colours, but to date most examples have been purchased by emergency and medical services, with a number being dispatched overseas for service with aid agencies.

Now, in the aftermath of the Gulf War, the Royal Saudi Arabian Government has placed the largest single order to date for 150 POD ambulances mounted on V8 One-Ten Defenders. Despite fierce competition from around the world, including Japan, the POD and Defender combination won hands down, with fast delivery dates being just as critical as price.

The POD is basically a GRP module which is slid onto the back of a suitable carrier vehicle and can be hooked into the vehicle's power supply for lighting and ventilation. Many PODs are used as mobile ambulances and communications vehicles, but more and more are being used as specialist treatment clinics.

When the POD clinic reaches it's operating area, it can be quickly demounted by fitting four low-tech jacks, raising the POD a few inches, driving the vehicle clear, then lowering the jacks. The entire operation can be carried out by one man if necessary, but using two or more pairs of unskilled hands speeds up the process.

While the medic attends to patients in the POD, the Land Rover can be travelling to other areas to collect more patients. In Third World countries in particular, where medical facilities are few and far between and personal transport is unheard of, this modus operandi is invaluable.

Just about any medical use from innoculation centres to dentistry clinics can easily be accommodated.

Now that trade restrictions are being eased on South Africa, a number of PODs have been ordered as remote area health clinics. In Japan, a mother and baby unit is doing the rounds, albeit on a Japanese base vehicle.

In Kuwait, the fire fighters extinguishing the oil wells have three PODs rigged for control and breathing apparatus recharging duties. On the overseas military side, an Australasian government has purchased POD for the Command and Communications role and a Far East government is about to place an order for a number of units for it's army.

PODs can be fitted to any vehicle from low-cost pick-ups to high mobility 4×4s, but the Hi-Cap One-Ten is widely regarded as being the best carrier vehicle. It is the one

By Bob Morrison

△ The POD is a demountable unit and can be kitted out for any number of civilian or military uses.

Kuwait

most recommended by their sales manager, ex-Royal Marine Steve Russell.

The Saudi POD ambulances use left hand drive V8 One-Ten Hi-Cap Defenders, complete with factory fitted cab air conditioning, as base vehicles. An umbilical cord from the Defender to the POD allows the ambulance body to be fully lit and air conditioned (roof-mounted pack) and a 240 volt electric supply is provided via an uprated alternator and Powergen unit.

When the POD is dismounted, it can be connected to an external 240 volt supply or a generator to maintain the lighting, air conditioning and electrical equipment. It is intended that these PODs will spend most, if not all, of their service lives attached to their parent vehicle, but the ability to demount the body gives the end user the option of using the Defenders for other duties in the future.

Internally, the POD can be configured in many ways to suit the end user. The Saudi vehicles, for example, were fitted with seating/stretcher rack on the left and a work-top module on the right.

Alternatively they could have been fitted with stretchers both sides or a combination of stretcher and Gurney. In addition, scoop and vacuum stretchers, blankets, oxygen, a resuscitator and a full range of first aid equipment are carried and a small built-in sink provides hand-washing facilities.

Externally, fold down rear steps give easy access through the tinted glazed double rear doors and a large tinted side window gives adequate natural light whilst retaining patient privacy.

Red Crescent markings are carried, but neither beacons or light bars are carried, although the manufacturer can supply either if required. The camouflage pattern is two-tone sand and pink similar to the mix which found favour with the British on Operation Granby.

At the time of writing, the first batches of PODs are already in the Gulf and the last batch should be well on it's way. Now, with such a large military order under it's belt, hopefully this successful civil concept will make major inroads in military markets. With it's ability to turn non-dedicated off-road vehicles quickly into dedicated high mobility ambulances, the POD system is a winner.

I'd like to express my thanks to Steve Russell at Shanning POD, Ashlyns Hall, Berkhamsted for his hospitality when I visited them during one of their most hectic periods.

● Finally would Stan Standley, formerly of 16/20 Squadron Detachment, Op. Granby, or anyone knowing his whereabouts, please contact me at LRO.

MILITARY SCENE

MILITARY SCENE

△ A Commando lightweight being deposited on the deck of RFA Argus for the sea voyage to Turkey and (inset) Land Rovers and helicopters travel as deck cargo on the Argus.

CHOSC Lightweight still in Gulf camouflage deep inside Northern Iraq.

△ *The vital role of providing a communications link back to the UK falls here to a 110 belonging to 30 Signals Regiment.*

Operation Haven

THIS MONTH we are going to take a look at the Operation which has received precious little Press coverage — Operation Haven.

The dust had hardly settled in Southern Iraq when the world suddenly became aware of the plight of the Kurdish refugees on the borders of Northern Iraq, Turkey and Iran. Within days, the international media circus was flying in with TV cameras instead of loaves of bread and moaning about how long it was taking the coalition to put troops on the ground to help the displaced Kurds fleeing from Saddam's bullies.

These media personalities were the same guys and gals who covered the Gulf War and forgot just how many weeks, not days, it took to get a creditable military presence to the Kuwait border.

In an incredibly short time, the RAF had actually dispatched Hercules transport aircraft, followed by Chinook helicopters to the theatre. Within hours of a political decision being taken by the Government to deploy British troops, the first Royal Marines of 3 Commando Brigade were on their way, and within a few days the first sizeable contingent of Land Rovers was heading for a Turkish sea port on board our old friend the *RFA Argus*.

At first it was not clear whether it would be the Paras or the Marines who would go, but in the end the Commandos won, mainly because they were equipped for high altitude, cold weather operations and had their own integral helicopter support. Regular readers of this column will twig from this that CHOSC were on the move again. I went down to Portsmouth to watch the Argus being loaded to the gunnels with helicopters, Bedfords and Land Rovers, but as CHOSC have had more than enough free publicity in LRO (no Andy — you're not getting your photo in this month) I concentrated on the Brigade Air Squadron's vehicles.

Although it was originally intended that no-one who had served on Operation Granby would be deployed to Northern Iraq on Operation Haven, in the end this aim proved to be impossible to achieve. Some of the CHOSC guys in particular only had a few days with their families before they were back in the field.

Others re-deployed to Turkey in support included RAF and Royal Signals units who had also seen Gulf service. Helicopters and equipment in transit back from Saudi Arabia were diverted straight into the operational zone.

By the end of April, the complete 3rd Commando Brigade and its support was in place and stability appeared to be returning to the region. However, media interest in the Kurds was by this time on the wane and coverage of the sterling work being carried out by the Brigade had all but dried up.

If a week is a long time in politics, it's an eternity to TV News producers and newspaper editors. Attention quickly moved from the Kurds to Bangladesh, then to pit bull terriers, then to the Gandi assassination, then the Ethiopian civil war and at time of writing the focus is now back on Saddam's boys hounding the Shi-ite Muslims near Basra.

As usual, the British Military is quietly getting on with the hearts and minds job — good news never makes the headlines.

Much as I would love to get out to the theatre of operations, it is not really a practical proposition. However, friends serving on Operation Haven have kept us supplied with Land Rover shots for which I am most grateful. I'd also like to thank DPR(N) and the Royal Marines for their help.

MILITARY SCENE

Photos from Iraq

MILITARY SCENE

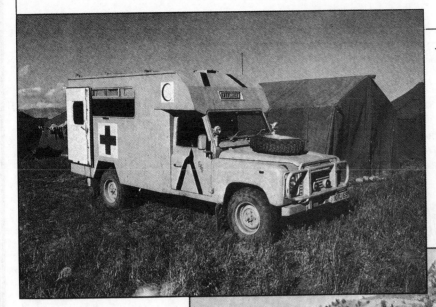

◁ The CHOSC 130 Ambulance still in Gulf War maskings.

Despite their age, the 101s still provide the Royal Marines with sterling service. ▷

Series III 4-stretcher ambulance — that's an Iraqi ◁ mountain behind, not Dartmoor.

The aid agencies are also working with the Kurdish refugees — this One-Ten belongs to Oxfam and the French charity Medicin San Frontieres also uses One-Tens. ▷

By Bob Morrison

Red on Green on Load gone

A 101 FC and 105mm gun are loaded . . . then they're gone ▽

◁ A text book landing for this Land Rover and trailer combination

AN empty hanger on an airfield, Lightweights, 109s and 101s are loaded onto pallets known as Medium Stressed Platforms. Each MSP can carry either two lightweights, a 109 & trailer or an FC101 and 105mm Light Gun. The vehicles are stripped down to windscreen base level, loaded with ammunition and stores, positioned nose-down on ramps on the MSPs, and then braced to absorb the impact with carefully located shock-absorbing wooden struts and packing. Working to a meticulously planned sequence, the Paras and RCT air dispatchers ready the MSPs for an eight hundred foot drop into 'enemy territory'.

A few hours later, the Land Rovers are fully rigged and being transported to waiting aircraft. At another location not far away, the men of the Leading Parachute Battalion Group prepare to don parachutes. The MSPs are loaded onto wheeled transporters for loading into the Hercules. First to be manhandled onto our aircraft is an FC101 of 7 Para Regiment Royal Horse Artillery. It is tied down, then a 109 and trailer is loaded into the hold behind it. The Parachutes for our FC101, loaded on the 109s MSP, are attached and the parachutes for the 109 are positioned on the tail ramp of the Herc. We are ready to roll.

The formation of Hercs, probably as many as fifteen in our wave, line up for take-off. This is only an excercise, so we take off in quick succession, breaking right and left alternately over the trees at the end of the runway. For the next couple of hours we will seldom rise above a hundred metres as we hug the ground contours to avoid radar. As the pilot banks from side to side to avoid over-flying villages, yours truly standing in the cockpit reaches for the first of many sick-bags.

Over the headphones I hear my cue - 10 MINUTES OUT. The loadmaster signals me down into the hold and I squeeze between the forward MSP and bulkhead. Hanging onto the urinal, I fasten the belt around my waist and check my strop.

The aircraft banks from side to side down the valleys and I lose all sense of direction. The tail opens up and I ready my camera. The loadmaster holds up one finger - behind the red light is on.

Suddenly the green light comes on, there is a bang as the first parachute deploys behind the Herc and jerks out the rear MSP. A second later, the forward MSP starts rolling and I hit the shutter release and start running.

LOAD GONE! I sit on the ramp and watch the following aircraft disgorge it's load. A couple of kilometres away, the Paras are dropping from the sky.

My thanks go to LXX Sqn RAF, 47 Dispatch Sqn RCT, 7 Para Regt RHA, and all the Paras and aircrew who made things possible.

By Bob Morrison

Portuguese Series III 86" hard top

Iberian co

MILITARY SCENE

THIS MONTH'S column is aimed primarily at military modellers who fancy building something a little different for their collections based on the Italeri Series III 109. Although, as readers of this column, modellers are in the minority, I know from my postbag that there are a fair number out there world-wide, who buy the magazine primarily for modelling inspiration. Hopefully the rest of you will be content with the pictures of unusual variants while I prattle on about cutting plastic for a while.

Whilst on an exercise with ACE Mobile Force at Santa Margarida in Portugal, I had the opportunity to photograph both Portuguese and Spanish variants on the Series III chassis. For the first time in AMF's thirty year history, both Spain and Portugal deployed artillery batteries for a live fire multi-national artillery exercise and both national contingents used a small number of Land Rovers in specialist roles.

In the past, Land Rover sold to many armed forces on it's reputation alone, but today politics is often the major decider of a military procurement contract. Any home grown vehicle which can provide local employment tends to stand a better chance of selection than a foreign built vehicle, even if it is not necessarily the best vehicle for the task.

For example, the Portuguese Army now uses the home-built UMM as it's light utility vehicle. This slab sided, steel bodied, leaf sprung 4×4 is in many ways reminiscent of the Series I long wheelbase and offers just about as much comfort. However, the Portuguese conscript is not likely to voice complaints too loudly! For range duties, a small number of American supplied M151 MUTTS are still in use, but they are not allowed out on the public highway as they are considered to be too dangerous.

Unlike the Army, the Portuguese Air Force still uses quantities of Series III long and short wheelbase hard tops. All such Land Rovers that I spotted, were fitted with winches, radios and box-section roof racks, with the 109s being used primarily in the communications role. The one-piece (GRP?) hard tops are unlike any that I've come across before and have a unique single panel, top hinged rear door. Both the 88" and the 109" have windows in the rear door, but only the 88s appeared to be fitted with rear side windows.

Judging by the small wing mirrors and the British Leyland mudflaps, I reckon they are pretty old, but I was not able to check chassis numbers to confirm their precise

△ *Santana Militar ambulance*

◁ *Portuguese Series III 109" hard tops*

nnections

age. Despite looking well used, particularly inside, they all seemed to be in very good condition. Of course chassis rust in the Portuguese climate is not a problem and Air Force vehicles are generally better treated than their Army counterparts.

One of the 88s was used as a general runabout by the Fire Section and seemed to be on the move every day from dawn to dusk. I followed it down rough tracks a couple of times in both a MUTT and a newer UMM, and it left both standing as neither gives the driver a particularly easy ride over rough ground. In fact in the back of the UMM I felt positively unsafe as each bump threatened to jolt me over the ridiculously low tail.

Modelling the Portuguese Series III should be quite easy if you have the confidence to scratchbuild a new hard top from plasticard. With a little bit of skill, it should be possible to form the main roof and sides from two rolled pieces, with a join level with the rear edge of the doors. The join can then be covered with a good layer of filler and sanded smooth when dry. Although at first glance the winch looks complicated, most of the detail is hidden by the front plate, and the rectangular control box mounted above the drum.

Portuguese Land Rovers are painted in a three colour camouflage scheme of broad vertical swathes up the sides and over the roof. The base colour is a light olive drab, very similar to well faded NATO green as used on British Series IIIs. The next colour, sprayed with an almost hard edge, is a dark green which is close to WWII Luftwaffe Dunklegrun.

The final colour appears to be charcoal black in some light conditions, but could well be a very dark shade of olive green. Weathering makes it very difficult to guess exactly what shade this colour was when fresh and it probably has anti infra red properties as well to confuse the eye even more. In addition Santa Margarida is a very dusty place and all Rovers had a thin coating which watered down the colours even more.

In contrast to the almost standard Portuguese Series IIIs, the Spanish Land Rovers were completely different. Built in southern Spain by Santana, the "Militar" range could best be described as akin to a long wheel-base version of the Series III lightweight.

Spain also has a policy of buying home produced military vehicles with latest purchases being of Spanish assembled Nissans, but a fair number of Land Rovers are still

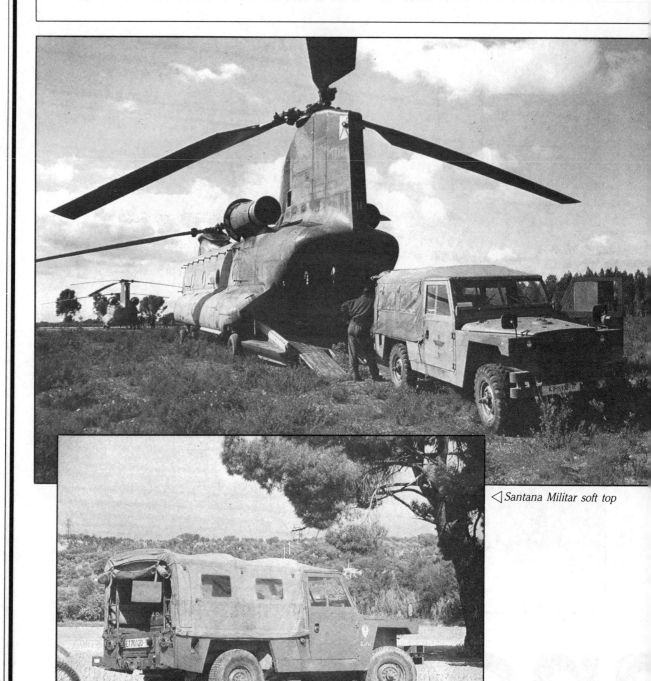

◁ *Santana Militar soft top*

in service and will be for some time to come. There is no longer a formal connection between Land Rover and Santana.

On this year's exercise, three versions of the Militar were fielded. The basic soft top version of the 109" chassis was used in the transport and cargo roles by both the artillery and aviation elements. A hard top FFR was used by the artillery in the communications role, and an ambulance version was on hand to provide first aid cover during parachuting and live firing.

If you fancy trying your hand at modelling a Santana Militar, the hard top is probably the best one to start off with, as, apart from the roof and bonnet, it is all flat panels. The chassis, wheels, bumpers, cab floor, load bed, bulkhead, windscreen and door tops can all be used, although the outer edges of the bulkhead wiill have to be pared back in line with the edges of the windscreen.

By careful trimming, the kit bonnet can be utilised, and even the kit canvas can be padded out with filler and sanded smooth to give a passable representation of a hard top roof.

Apart from the reduced width, all major dimensions are similar to the standard Series III and it should not be too difficult for anyone with a modicum of scratch-building experience. One minor point of note — the hard top photographed had an extended radiator grille, but all other Santanas that I've seen have had the more simple mesh arrangement.

Building a soft top Santana will be a little more difficult than building a hard top as the tilt and tilt frame are considerably different to the Solihull version. The body sides, rear and tail-gate on the Spanish vehicle are much lower than the standard Series III; tubular guard rails, which rise level with the bottom of the door windows, being fitted for passenger safety when the tilt sides are raised.

However, like the hard top, the basic body panels are pretty straightforward. Rear lights and wing mounted convoy lamps are of the American style and can be poached off an old Jeep kit or similar.

The four-stretcher ambulance will be the most difficult of the three versions to build, but an experienced scratch-builder should be able to turn out a smart addition to his miniature Land Rover fleet with a little patience. The cab and bonnet arrangement is identical to the basic vehicle, and the rear

△ Santana Militar hard top

body is constructed along relatively simple lines.

Only the roof will pose real problems, and ideally this should be heat formed over a balsa mould, but it is possible to laminate it if your vac-forming skills are not up to much. The body sides are basically flat, with a simple curve behind the doors, but note how the panels over the door narrow in to line up with the edges of the windscreen. The gutter and raised sheet joints can be fashioned from microstrip.

Internally, the seating and stretcher racks are perfectly straightforward.

Although the Santana Land Rovers were all painted in the same shade of dark olive, the bleaching effect of sun and a fine coat of dust made them look more brown than green in certain light conditions. Number plates consisted of black letters and digits in white painted panels, with the characters having a distinctly Spanish style.

Footnote

Regular readers of this column will be aware that for the past thirty years, ACE Mobile Force has undertaken the role of NATO's premier deterrent force against infringement of any member countries' borders by the Warsaw Pact, particularly on the remote flanks of Europe. Because AMF has never had to deploy operationally, many see it as having been ineffective, when in reality by never having to deploy it has done the very thing it was created for — it kept the peace!

In the new post-Iron Curtain Europe, the NATO AMF concept is being expanded as conventional armies contract, and the Force's future seems assured. Most European countries already contributing are increasing their commitment to AMF as they decrease the numbers of conventional troops in central Europe and others like Spain and Portugal are now very keen to play a major part in joint European security.

It is even possible that the emerging democracies of Eastern Europe, not to mention the Soviet Union, may take part in a pan-European defence force in the not too distant future. It may sound far-fetched, but it is less than two years ago that I looked over the German border at grim-faced, machine-gun toting, communist regime guards.

Last month I drove from Ostende to Prague without having to get out of my Land Rover at a single border crossing.

MILITARY SCENE

△ Land Rover leaving the Hercules at 800 feet.

Falling from the sky

Dear Defence Secretary,
Under "Options for Change" you promised that the leaner British Army would have the best equipment available to help it defend Britain and the Free World's interest. How about some new Land Rovers for the Paras then?
Yours ever,
Bob.

SERIOUSLY though folks, the lads really could do with some new kit as their Forward Control 101s and Series IIIs are getting just a little bit knackered from being thrown out of the back of aircraft on a regular basis. Even the poor old Land Rover occasionally objects to hitting Mother Earth after an eight hundred food fall.

Over the years, the technique of packing and strapping a 109 and trailer or a 101 and Light Gun to a Medium Stressed Platform to enable them to be airdropped has been perfected to such a degree that even when things go wrong, little real damage is done.

Earlier this summer, I watched a Series III 109 come down on 1½ of it's 4 chutes after a glitch as it left the Hercules. As it hit the ground a couple of hundred yards away, before the Land Rover that left the aircraft ahead of it, I swear I felt the thud. When it's owners got to it they were expecting the worst, but although at first they could not even get it to turn over, when the driver bump started it off the MSP it roared into life and drove away.

Talking to the experts later, they reckoned that the shocks had been subject to 6 inches of instant compression. The only visible external damage was a slightly rippled front wing, but the bump stops looked a little distressed.

Some of you might have spotted a piece in one of the "respectable" newspapers a couple of months back decrying the One-Ten and claiming that it could not be airdropped as the chassis was too weak. As always, the newspapers don't give the full story. The One-Ten has been dropped, mainly by the SAS not the Paras, and their vehicles are to a different payload specification to the run of the mill Land Rover.

Although it is believed that some weak points have been identified on the chassis, this may well be to do with the method of support used or any one of a dozen variables. The way that any vehicle is

◁ *The RAF having put this Series III into the trees, the Paras have to get it out.*

Few vehicles are worked harder than Para 101 Gun Tractors. ▽

△ *Believe it or not, this Series III landed with only one and a half of its four chutes deployed, but drove away.*

cushioned from impact is under constant review to find the ideal set-up. It is only after extensive simulation followed by a number of operational drops in varying conditions, that any vehicle is declared 100 per cent droppable.

No vehicle is undroppable — it may just need a little more investigation to find a solution to it's problems. Even after more than two decades of dropping Series II and III Land Rovers, the occasional problem can still be encountered in the field.

Incidentally, vehicles dropped are not modified specially for the task, and even your civilian Land Rover could be rigged for an airdrop with only a few hours preparation.

Now that the One-Ten has been declared fully operational for Royal Marine use and the first Commando is about to convert, hopefully it won't be too long before the Para battalions are re-equipped. However the question of what to replace the 101 Gun Tractors with has still to be answered.

I personally don't believe that the RB44, the only other vehicle in this category currently in service, will fit the bill any more than the now defunct Land Rover Llama would have. The cost of developing a completely new vehicle just to equip a few specialist artillery batteries would be prohibitive, but a special body on a One-Ten or One-Thirty chassis might be the answer.

With the reluctance to use turbo diesels disappearing at last, the new Tdi engine could be just what is needed. The One-Thirty is, of course, in British and foreign service as both a Rapier Tractor and Airfield Ambulance, albeit only in V8 form at present, with at least one having been well trialled by the Royal Navy in the Gulf War and in Turkey.

As you read this I should be somewhere on the continent on exercise with either the multinational ACE Mobile Force or with units dedicated to NATO's new Allied Rapid Reaction Corps. Having gone soft recently over the Discovery, I felt it was time to go back to my roots. I'll be following the action around Germany and Denmark in a Defender Ninety, but naturally I've requested a Tdi version from the factory.

Sign of the times? An officer friend serving in Germany was moaning that his NCOs are all running around in Discoverys which are replacing their GTi's and XR3's as crumpet catchers. We were born in the wrong generation weren't we Brannigan!

Bob Morrison

MILITARY SCENE

MILITARY SCENE

◁ REME Light Aid Detachment lightweight

One Ten hardtop FFR ▷

Civilian pattern Royal Yeomanry One Ten ◁ Station Wagon

Although most Yeomanry Land Rovers are now coil-sprung, a few Series IIIs still soldier on ▷

△ Series III Unit Ambulance

Royal Yeomanry

IT'S A Friday evening early in September somewhere in England or Northern Ireland. Engineers, plumbers, accountants, sales reps, van drivers and a host of other tradesmen and professionals swap their work clothes for the uniformity of DPM combats. The Territorial Army, or to be more precise the Royal Yeomanry, is on the move.

From Belfast to Croydon and Nottingham to Swindon, Land Rover engines cough into life as the senior Territorial Army regiment of the Royal Armoured Corps departs for Germany in support of the Rhine Army.

Despite being manned by part timers, the Royal Yeomanry is equipped to the same standard as a regular unit, and at present is larger than any other regiment in the Royal Armoured Corps, As it is primarily an Armoured Reconnaissance Regiment, it is equipped with Fox armoured cars and Spartan armoured personnel carriers, but also numbers a varied collection of Land Rovers on it's inventory.

When their press officer issued an invitation to visit the Royal Yeomanry in the field during exercise Certain Shield '91, how could I refuse?

By Sunday morning, the entire regiment, comprising headquarters and four squadrons, with attached Military Police, had arrived at their holding area on Sennelager, and final equipment preparations were underway. The Royal Yeomanry were to act as enemy "Gold" reconnaissance forces over the next fortnight in an exercise that would test the new NATO Multinational Northern Army Group Airmobile Division (MNAD).

As the armourers fitted weapons to the vehicles and the REME Light Aid Detachment fixed the odd mechanical defect, the old hands ran the clutch of new recruits through weapons drills for the umpteenth time. Working alongside Dutch regulars, they would soon be pitting themselves against a formidable attacking force and the Royal Yeomanry were determined that their regiment would be no push-over.

Some TA units have a high turnover of personnel as young guys (and girls) find that family commitments leave less time for the regiment, or they lose the buzz as training routines and exercises in the UK become a bit of a bore. Due to the varied and interesting role of the regiment, the regular trips abroad and the high degree of mechanisation, the Royal Yeomanry does not suffer as much as some TA units, and tends to keep it's personnel longer than most.

One noticeable sign of this is the age and evident experience of the NCOs, most of whom could easily pass for regulars, and the professional attitude of the ordinary soldiers.

The Yeomanry tradition goes back nearly two centuries to the days when France declared war on the United King-

LRO stickers get everywhere ▷

These Military Police NCOs, from 153 Provost Company working alongside the Royal Yeomanry, are also Territorial Army soldiers ▽

dom and Parliament authorised the formation of volunteer cavalry. In the early days, the Yeomanry troops were recruited from the residents of rural towns or the employees of country estates and were paid as regulars when called out by the sheriff or Lord Lieutenant.

They initially had to provide their own horses and arms but in time the government provided funding. By the latter half of the 19th century, most Yeomanry Troops were integrated into their county's regiment.

The present Royal Yeomanry squadrons carry the names and traditions of some of the oldest and most famous of the old Yeomanry formations.

Of the other Yeomanry regiments in the British Army, only the Queens Own Yeomanry operates in the Armoured Reconnaissance role in support of BAOR.

The Royal Wessex, the Queens Own Mercian and the Duke of Lancasters Own all operate in the Home Defence role and use civilian pattern Land Rovers in place of armoured cars and armoured personnel carriers.

On the Land Rover side, the Royal Yeomanry is equipped mainly with One-Tens and Nineties, but they also use a few Series IIIs. Their medical section runs a Series III ambulance, the Light Aid Detachment operates a soft-top lightweight and there is even a civilian pattern station wagon on strength.

This naturally aspirated diesel One-Ten has been used on several NATO exercises, as well as for routine duties, by the RSM and press officer, who have nothing but praise for it.

Under "Options for Change" and the shrinking British Army presence on the continent, it looks likely that the Royal Yeomanry may lose it's armoured vehicles and re-equip with Land Rovers.

Full details have yet to be announced as I write this, but it seems likely that the regiment will survive the current round of amalgamations, though it may lose at least one of it's squadrons.

Despite this, morale is high, and the traditions of old Yeomanry units like the Sherwood Rangers or the Kent and Sharpshooters will live on. They have been through amalgamations before and survived.

When I left the Royal Yeomanry, they were preparing to move out to form a reconnaissance screen ahead of the main force.

Next month we will look at one of the Brigades they successfully kept tabs on - Britain's 24 Airmobile.

To get me around Germany and Denmark for this year's major exercises, I borrowed a hard top Defender Ninety Tdi from the factory.

This was my first long distance test of the new engine in a Ninety, and I was very impressed with it.

Moving the goalposts

By Bob Morrison

FINAL tenders for the Ministry of [Defe]nce TUL/TUM contract have gone in [and] everything should now quieten down [for] a couple of months until the competi[tive] trials stage commences.

[In] the end, only three companies seem to [have] tendered, although there are rumours [of a] fourth bid from an aerospace compa[ny]. The confirmed manufacturers are Land [Ro]ver, Steyr-Daimler-Puch and Iveco Ford. [W]hull still won't pass comment on exact[ly w]hich vehicles they have offered, but I [bel]ieve all three will be Defender based, [with] the Discovery pick-up being a red her[ring].

[S]teyr-Daimler-Puch are offering short [and] standard wheelbase versions of the [ir] Pinzgauer for the TUL and TUM plus a [6x6] version for TUM[HD] - the existence of [the] short variant was kept under wraps [unt]il the tenders went in. Iveco-Ford are [offe]ring a military version of the Turbo [Dai]ly, similar to that used by Italian Alpini [troo]ps, for TUM[HD] only. Interestingly, all [ma]nufacturers appear to be offering turbo [die]sel engines as standard.

[P]ortugal's UMM, considered by many [inc]luding myself as a likely front runner, [wit]hdrew at the last moment. It is thought [tha]t the high cost of providing up to eigh[tee]n trials vehicles and back-up with no [gua]ranteed return, was possibly the decid[ing] factor for this relatively small manufac[tur]er.

[E]veryone expected the G-wagen from [Me]rcedes-Benz to go through to the trials [sta]ge, but they also withdrew at the last [min]ute - less than a week before the tender [as] their sales staff at the Eurosatory Mili[tar]y Equipment Exhibition could not con[fir]m or deny rumours of the withdrawal. [Ge]rman politics was probably the decid[ing] factor in this case as their government [is p]laying down overt military sales whilst [che]mical and engineering expertise [app]ears to be covertly sold to all comers. [Ma]ybe if UMM had known that the G-[wa]gen was being withdrawn and that there [we]re only two other contenders offering [a] full range, they might have been tempt[ed] to stay the course.

[T]ender documents for two other MoD [req]uirements are also due out shortly. The [firs]t is for around 300 4x4/4x2 pick-up [tru]cks offering limited off-road capability, [on]e tonne payload and a three man cab [ext]endable to five. The other requirement [is f]or a fleet of 500-800 all wheel drive, [me]dium mobility, four stretcher battlefield [am]bulances.

[A]lthough Land Rover can supply a [De]fender variant which meets the pick-up [spe]cification, MoD is probably more inter[est]ed in a mass-production commercial [tru]ck. However Land Rover should be a [pri]me contender for the ambulance com[pet]ition, either in their own right or as a [sup]plier of base vehicles to the three spe[cia]list coach builders also on the list.

[T]he new ambulances will eventually [rep]lace the well used fleet of 16 year old [1]01s and the earlier Series II & III 109s [sti]ll in widespread use as unit ambulances. [So]me 25 year old Series II ambulances can still be seen in daily service, which explains why the tender summary states from the outset that the new vehicles will have a planned life of at least 15 years.

Incidentally, the Defender battlefield ambulance shown recently at Eurosatory in Paris would have been the next French military ambulance if the specification had not been subsequently re-written to suit a van-derived vehicle. It's amazing how often goalposts move in military procurement.

Whilst on the subject of military ambulances, no doubt many of you read Paddy Ashdown's attack in the press of 24 Airmobile's ambulances in Croatia. He was quite right in his criticism, as these sixteen year old vehicles are right at the end of their service life and were originally supposed to be left in the Gulf at the end of the war.

However, as always, the politicans only tell part of the story. At around the same time, a senior MoD source confided in me "off the record" that a fleet of virtually brand new Defender 130 airfield ambulances had been prepared by the army for dispatch to Croatia weeks before, but bureaucracy and indecision over funding of their transportation had stranded them at depot.

Unlike others, who supply brand new equipment to the UN then knock the bill off next year's contribution, Britain plays by the rules. Consequently our guys have to lay their lives and those of their patients on the line in clapped-out mid-seventies vintage vehicles, while the civil servants and bureaucrats go through the proper channels to keep the bookwork in order.

Going back to Eurosatory, also on show was the new CAV 100 lightweight armoured Defender. This is not the place to pass more than brief comment on the new vehicle's properties save to say that it offers significant additional crew protection for internal security and peace-keeping duties.

The CAV 100 Vehicle Protection System uses S2 Glass composite bonded material in a modular monocoque construction assembly tailored to provide maximum ballistic protection.

A significant order for these vehicles has been placed by the MoD; probably to replace ageing VPK Series IIIs.

▽ *French military pattern Defender ambulance at Eurosatory, Paris*

by Bob Morrison

Lightweights and One Tens

OVER the last couple of months, I've had a few phone calls from military enthusiasts who have spotted fleeting glimpses of what appeared to be lightweights on satellite newsreel footage coming out of Yugoslavia. Some callers thought that they might be Santanas as they are slightly different from British military spec, but I suspected that they were probably decommissioned Dutch vehicles. Earlier this year a large number of reasonably priced lightweights appeared on the Dutch civil market as their Army fleet went Austro-German.

Confirmation that these vehicles were Solihull-built Dutch Army machines finally came from combat photographer Yves Derbay when I managed to speak to him on one of his brief visits home to Paris between assignments. It appears that the Croat Army managed to purchase a quantity of these vehicles soon after declaring independence last June and pressed them into service immediately. If he gets the chance, Yves has promised to photograph one being used by it's new owners, for LRO.

Dutch One-Tens

Although the Dutch Army has made the sad mistake of going continental in it's choice of Land Rover replacement, the Royal Netherlands Marine Corps has more sense. They have now been using the One-Ten for more than four years and seem very happy with their choice. During the autumn exercise season I bumped into them on several occasions both in the UK and abroad.

Whenever possible, I took the opportunity to ask their drivers for a personal view of the One-Ten. Without exception, every driver I spoke to had nothing but praise for his machine. The only adverse comment that I heard on the Dutch vehicles was that most users would have preferred an insulated hard top for cold weather work.

Incidentally, unlike the pony-tailed, ear-ring-wearing, Dutch Army, all Netherlands Marines (even conscripts) are volunteers and are rated almost as highly as our Royal Marines with whom they work in close contact.

In Denmark, on an ACE Mobile Force deployment, I also spotted a brand new Defender One-Ten in Dutch military service. This Land Rover was attached to the Force Artillery HQ in support of the Dutch heavy mortar battery. At first glance I thought this was just another of the original batch of One-Tens used by the Dutch Marines, as it had all the minor external modifications of RNLMC specification Land Rovers, but when I glanced at the rear I noticed the Defender logo beneath the national tri-colour.

Sure enough, when I looked closer at the bonnet, it also sported a Defender nameplate. As is par for the course, Solihull declined to comment on the quantities of Defenders purchased by the Dutch military, but they were able to confirm that there have been a number of follow-up orders since the RNLMC re-equipped with the One-Ten.

Stop Press!

Two bits of procurement gossip just in. The JRA 6x6 was one of the early contenders for a Canadian Defence Force replacement vehicle programme, but it appears that an Italian vehicle has won the "contest". Having said that, there is no money in the coffers as yet for the order to be placed – Canada is suffering even worse than the UK in the current world recession.

Interestingly, this competition was purely a paper affair, with no competitive comparison or end-user trials being undertaken. I understand one of the shortlisted vehicles was not even in production at the time of tendering!

Bearing in mind that the Italian vehicle was well and truly trounced by the One-Thirty during RAF trials, it will be interesting to see if this method of vehicle selection is used next time round.

Back on this side of the pond, the Ministry of Defence has now issued the Invitations to Tender for the TUL/TUM (Truck Utility Light/ Truck Utility Medium) vehicle replacements programme. Almost thirty companies have expressed interest in tendering, so Solihull will really have it's work cut out to win this one. French, German, Italian, Portuguese and North American manufacturers will be using every trick in the book to land this contract and it is more than likely that European-assembled Japanese vehicles will also enter the fray.

A full evaluation programme should sort out most of the wheat from the chaff, but Land Rover will have to kick chivalry out of the window and go for the throat like the continentals, if it aims to win. I'll keep you posted on anything I hear.

Apologies. Due to lack of space and my unfortunate inability to work to a brief (well someone has to give the Editor a hard time) the promised feature on 24 Airmobile Brigade in Germany has had to be held over. Also in the pipeline – ACE Mobile Force in Denmark.

The RNLMC use One-Tens to tow 120mm heavy mortars.

△ Dutch One-Ten fords at speed
The Dutch are buying Defenders too ▷

MILITARY SCENE

Top: Twenty one year old ambulance in use. Above left: Dutch Marine One Ten. Above right: AMF Lightweight and Ninety

Ace Mobile

LAST SEPTEMBER I had the pleasure of spending some time in Denmark with the multinational ACE Mobile Force. Unlike my previous trips with AMF to cover Ardent Ground 90 in Germany (LRO July 90) and Ardent Ground 91 in Portugal (LRO August 91) which only involved the Force Artillery, Action Express 91 was a full Northern flank deployment. British, American, German, Belgian, Italian, Dutch and Luxembourg troops took part alongside Danish Army and Home Guard troops.

Had the Warsaw Pact Baltic Fleet gone to war, only little Denmark and neutral Sweden would have stood between them and the oceans of the world. With Southern Fleet access to the free world being even more restricted through the Bosphorous, and ice blockading the northern Soviet ports during winter, one of the first priorities of war would have been to occupy Denmark. This of course is exactly what Germany did during the Second World War. AMF's task was to convince the Soviets, or anyone else for that matter, that Denmark's NATO allies were both willing and capable of coming to her assistance.

Action Express 91 was designed to allow AMF to practice deployment and simulated defensive combat whilst giving the Danish Army and Home Guard the opportunity to train for the defence of their homeland. AMF joined the Blue force of 2 Zealand Brigade, 2 Zealand Combat Group and the Danish Home guard in the defensive role whilst 1 Zealand Brigade, the Zealand Recce Battalion, the Bonholm Combat Group, a German Airborne Battalion and 23 SAS played the part of attackers.

Airborne, amphibious and simulated Spetnaz (23 SAS) raids were mounted against Blue Forces tasked with the defence of Zealand, Falster and Lolland islands. During the initial Deploy and Deter phase, the artillery units carried out a series of live fire exercises under the code-name Action Barbara at Jaegerspris ranges to the north of Zealand.

A newcomer to AMF Danish deployments was 2 MARBAT of the Royal Netherlands Marine Corps. As regular readers will know, the Cloggies are confirmed users of the One-Ten and naturally their Land Rovers featured prominently in Action Barbara. As I've covered the Dutch Marine One-Tens on several occasions, I won't dwell too long on the subject this time, but it was interesting to see their heavy mortar battery using Land Rovers to tow their 120mm mortars.

Although the British Contingent had not been re-equipped with winterised Defenders in time for Action Express, it was noticeable that the Lightweights and 109s were slowly being swapped for Nineties and One-Tens. The Red Hand Gang of 244 Sig-

Report and photos: Bob Morrison

AMF Commander General Carstens with his Series III 109 FR ▷

Red Hand Gang 110 supporting the force helicopter unit ▽

Gurkha signaller readies his 109 for transmission ▷

Force

nals Squadron (Air Support), who provide air liaison communications were trogging around in One-Tens this time, but poor General Carstens had to make do with a hard top Series III FFR.

By the way, I'd like to take this opportunity publicly to thank General Carstens for his help over the last couple of years and wish him luck in his new command.

Another Land Rover variant still in service with AMF long past it's sell by date is the 109" field ambulance. The one in the accompanying photograph is no less than 21 years old. Rumour has it that the Ministry of Defence is soon to issue a requirement for replacements for these vehicles, some of which are over 25 years old but still in regular service.

The One-Thirty, currently used by both the RAF and the Royal Navy as a fast response airfield ambulance, would be ideal. When the RAF One-Thirty ambu- lances were ordered, the V8 was the best off-the-shelf engine available for high speed, but today the Tdi would probably fit the bill.

Talking of the Tdi, for the autumn exercise season, I took a factory demonstrator Ninety Tdi to Germany and Denmark. Leaving Devon on a Friday evening, I drove at high speed to Dover for the night sailing to Ostende. Next morning I drove across Belgium, through southern Holland and on to Sennelager near Paderborn in Germany, arriving there in time for Saturday lunch.

I drove extensively around the exercise area, often two or three hundred miles a day, on back roads, lanes and farm tracks until midday Wednesday, when I set off for a high speed run to the centre of Zealand in Denmark. I checked into the Allied Press Information Centre at Ringsted in the early evening.

Thursday morning saw me up at the crack of dawn to drive up to the Jaegerspris range in the north of Zealand, where I spent the next two days off-road or on rough tracks.

The girls flew out to Copenhagen to join me for the weekend, but come Sunday the Ninety was back on the road as AMF deployed south to Lolland and Faster.

Monday to Wednesday was spent in the field covering on average a hundred miles a day with the Danish, British and German recce troops – a fair bit of this mileage was off-road. After covering a bridge attack at noon on Wednesday, I headed back for Germany again, arriving at Sennelager early evening. Thursday was spent with the Royal Yeomanry, but as the exercise had been called off a day earlier than planned due to deteriorating weather, I headed homewards that afternoon.

It was on the return journey to Ostende, that I hit the one and only problem with the

MILITARY SCENE

△ Morrison's Ninety Tdi in the thick of the action as Danish orange forces move forward

◁ AA patrolman Maurice, a keen LRO reader

Tdi. At 90mph, in the outside lane to pass some slow moving trucks, I suddenly lost turbo power.

The poor guy in the BMW on my tail got the fright of his life when my speed dropped to 65mph instantly. Naturally I pulled onto the hard shoulder when safe to do so and dived under the bonnet with a torch.

Why do vehicles always breakdown a) in the wet b) in the dark and c) just after the only motorway exit for 20 miles?

As I'm primarily a snapper and scribbler, not a mechanic, I don't have much of a clue when things go wrong. I could see that there was oil everywhere but the dipstick was in place and there was no steady trickle from the engine, so I assumed (correctly) that it had to be a loose hose or something.

There was still some oil showing on the dipstick and the red light hadn't come on, so I chucked in some Superlube from the tool box and set off carefully for the port. Even if I'd found a local garage open that late in the evening, the chances of them knowing any more about Land Rover turbo diesels than me was unlikely.

By the time I got to the ferry, there was still some oil left and the red light was still off. After four hours on the ferry, the sawdust that I asked the crew to spread under the engine was virtually unstained, so I drove the vehicle off through customs, then called out the man from the AA.

When Maurice turned up in his naturally aspirated Ninety, he burst out laughing. He'd been killing time in the small hours reading back issues of LRO in his cab – honest.

He reckoned it was only a blowing hose connection as well, but as one of the cylinders sounded a bit unusual, he recommended a relay home. He was right – when Land Rover's mechanics went to work under the bonnet they confirmed that all that was wrong was a loose hose. Still when it's not your vehicle you don't want to take chances.

In all I did almost 3,000 miles in the Tdi in less than 2 weeks, returning on average around the 24 mpg mark. As well as the vast amount of high speed road work and general running about, I also ran cross-country with German recce units using Wiesel tracked vehicles. The lightweight Wiesels which run on rubber band tracks are incredibly fast and manoeuvrable, even over claggy Danish water-logged soil.

Interestingly, the Danish G-wagens chasing the Wiesels tended to take the long way around when they encountered ditch and bank obstacles. The Ninety went everywhere that the Wiesels went however, although not quite as quickly, of course.

▲ 5a Spanish 109 MILITAR in Portugal for a NATO multinational exercise.
▼ 5b Well used Danish Air Force Series III Station Wagon.

▲ 6a Special Forces line up of Arab Remote Area Patrol Vehicle, SAS Desert Patrol Vehicle and Australian 6x6 Long Range Patrol Vehicle.

▼ 6b The Special Operations Vehicle - a senior US officer has now publicly acknowledged that the Rangers are the end user.

▲ 7a Gulf War REME Light Aid detachment 110 armed with GPMG and toting a capstan winch.
▼ 7b The Classic Series III SAS Pink Panther.

▲ 8a Series III 88" Lightweight of 3 Commando Brigade on exercise ROLLING DEEP 92.
▼ 8b Royal Marine lightweight 88, rigged for heli-lift comes ashore from a landing craft in Norway.

Danish Military Land Rovers

Although the Danes have replaced most of their Series III Land Rover fleet with G-wagens, they still use a few in both army and air force service.

These photos are a round up of various types spotted in use during Action Express.

Above Right:
Danish Air Force Station Wagon used as an airfield security vehicle.

Right: All Danish Army soft tops seem to carry a roof rack like this one.

◁ *88" Army lightweight – the only difference between this vehicle and a British Army Lightweight is the night convoy light on the bumper and the tilt side windows.*

Army 88" soft top in use as an umpire's vehicle – note the registration number repeated on the wing ▷

MILITARY SCENE

Above: 24 Airmobile Brigade soft-top 90 and trailer. Below: 90 Gunship armed with anti-tank recoilless rifle

Bob Morrison

The perfect choice

△ *Medical 90 on Operation Desert Sabre — The Gulf War*

THE MINISTRY of Defence recently announced invitations to manufacturers to tender for the replacement of more than 8,000 General Service load, personnel carrying and variant vehicles which are nearing the end of their active service life. The requirement calls for commercially based vehicles in three payload classes between 500kg and 1400kg, preferably with a common basis. A further requirement for medium mobility 4x4 ambulance replacements has also now been issued. Over the next few issues, we'll take a brief look at the possible Land Rover contenders, starting with the 90.

TUL (Truck Utility Light) requires a vehicle capable of carrying 500kg powered by a diesel and AVTUR engine with a power to weight ratio of at least 20kW/tonne. Operating temperature range is to be between -31°C and +44°C and the chosen vehicle will be suitable for airportability and air drops. Maximum allowable dimensions are 3900mm long by 1800mm wide by 2300mm high. For operational reasons the vehicle should be capable of being stripped down to the level of the front windscreen base and be capable of accepting role changing applique kits.

The Defender 90 powered by either a naturally aspirated or turbo charged 2500cc diesel engine easily falls within the design envelope. Some commentators have suggested that the specification was written around the Land Rover, but even a cursory glance at the capabilities of the 90 disprove this. With a naturally aspirated engine the power to weight ratio is 33 per cent above requirement, the turbo diesel provides at least 35kW/tonne and the available load capacity of 900kg is 80 per cent greater than needed.

Weighing in at around 2500kg fully loaded, the Defender 90 is readily airportable by Sea King and Puma, it can be carried inside a Chinook or Hercules, and is capable of being dropped by parachute on a prepared platform (MSP). Loads of 500kg can safely be towed off road and 4000kg on road. Four men and kit can be comfortably carried in the rear, in addition to the driver and front seat passenger, and 920kg of cargo can be accommodated without removing the inward facing rear seating.

Normally, in the communications role (FFR), the rear compartment would carry radio equipment, batteries and two operators. Hard and soft top versions are in service with the British Army and overseas armies use gun ship variants. In the field, the 90 is regularly stripped down to windscreen base level both for transportation and to provide a low silhouette.

One of the greatest advantages of the 90 over it's rivals is its simplicity of construction and use of aluminium alloy body panels. By the very nature of their duties, Land Rovers tend to get knocked about a fair bit. Most of the competition has conventional welded steel bodywork which crumples in a shunt and requires extensive workshop facilities to bring it back to parade ground appearance.

Often when a Land Rover gets damaged in the field, panels can be easily hammered back into shape and made to look like new with a fresh coat of paint. Where damage is extensive, the offending panels are easily unbolted for replacement in minutes rather than days. Unlike steel body panels, the Land Rover's do not need immediate extensive paint treatment to prevent corrosion.

If short of specific spares, as happened to at least one unit I covered in the build-up to the liberation of Kuwait, Land Rover's world-wide support network is always just a phone call away. The company guarantees to get a part to any dealer on the planet within three days.

Already combat proven by all three British services in the 1991 Gulf War, the 90 is readily adaptable to all current roles with minimum preparation. It is in daily use by the British Army in all theatres as a load carrying, personnel, communications and liaison vehicle, and around seventy other nations employ variants of the 90/110 family for their military and para-military forces.

At the time of writing, the 90 is in service with the British Army from arctic Norway to the Falklands and from Canada to Hong Kong.

The Royal Navy carries 90s on board some of it's vessels and the type is to be found on just about every RAF aifield in the world.

MILITARY SCENE

△ *William Trump's wonderfully original Forward Control, now offered for sale to enthusiasts only*

FORWARD CONTROLS

An original

A FEW years back, farmer William Trump got the fright of his life when a car handbrake-turned into the filling station forecourt as he topped up his Land Rover and a gibbering photographer jumped out with camera on full auto. The object of my desire? A genuine pre-suffix registered 109" Forward Control in almost perfect condition for a working vehicle.

Although I didn't know it at the time, for the next year or so our paths would cross weekly as I headed east out of Devon on my way to work at the same time as William headed west to market in Exeter, usually with a load of pigs on board. Occasionally, we would both stop off at that same filling station for petrol and have a chat about 695 PTA.

During one such chat, William mentioned that the rear cross member needed replacing and asked if it was possible to buy a genuine item, but after much telephoning round Solihull and other contacts we drew a blank. In the end, it was decided that a military pattern rear cross member was probably the closest alternative.

However, as William was rather keen to keep the vehicle as near original as possible, he asked his local garage to handmake an accurate copy instead. As this is rural Devon, their mechanics are craftsmen in oily overalls, not vehicle servicing operatives in white coats, and they were only too happy to oblige.

William first saw Land Rover's new Forward Control model at an agricultural show in 1963 and was so impressed by it's potential as an all purpose farm vehicle that he immediately placed an order for one. Twenty-nine years later, after what amounts only to routine maintenance, the Land Rover is still in daily use around the farm and makes the occasional foray into market.

As this is a working vehicle, it is naturally not in concours condition, but to get it to this state would not take too much work. Only the tyres, one wheel and a few minor mechanical parts have been replaced over the years, which just goes to show what a canny choice William made.

Current mileage is under 30,000 and genuine, although William reckons you can add on another 29,000 for reverse mileage. In the days when they mixed their own feed on the farm, they used to daily reverse the Rover up a long, steep, muck covered ramp to collect the feed.

Amongst the vehicle's paperwork, which includes the original Owner's Manual published in October 1962, Maintenance Schedule, Service Guide & Warranty and Instruction Manual, are a couple of interesting letters. The first, dated 16th September 1963, is from the Technical Service Department at Solihull and refers to the replacement of a fractured rubber servo brake hose.

As a result of their investigation into the failure of this hose on William's vehicle, they altered the material specification and produced a new hose.

The second is from the Goods Vehicle Centre in Swansea returning a test fee. Yes, you read that correct, Swansea actually

By Bob Morrison

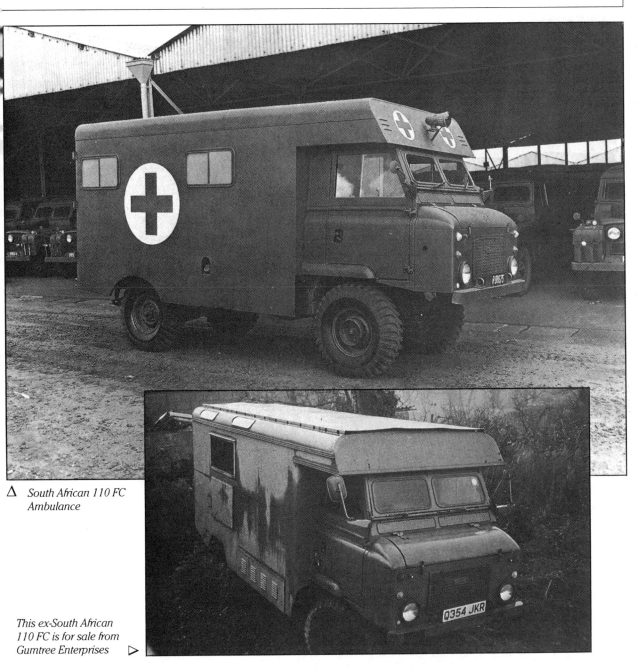

△ South African 110 FC Ambulance

This ex-South African 110 FC is for sale from Gumtree Enterprises ▷

once returned a vehicle test fee. As this document could well be of importance to other FC 109 users, we are quoting the relevant part in full with William's permission:–

The Goods Vehicles (Plating & Testing) Regulations 1968.

I refer to your letter of 17th June 1969 confirming that your Land Rover, registration number 695 PTA complies with Regulation 3 (1) of the Construction and Use Regulations 1966 and as such can be classed as a dual-purpose vehicle. Your test fee of £5 is therefore refunded herewith.

Current Livestock Regulations now preclude 695 PTA from being used to transport animals (she can be fitted with a rudimentary ramp to assist loading of same) and William no longer needs to use her for heavy duties around the farm, so her days at Strete Raleigh are numbered. Rather than letting her go to someone who might strip her down for logging work or similar and run her into the ground, William would prefer to sell her to an enthusiast looking for a near original vehicle to restore.

Being a low mileage regular runner in need of very little work, she can command a higher than average Land Rover price even amongst the local builders and contractors, but as a little piece of Solihull history, she should go for a song. The 109" Forward Control model was only produced in relatively small quantities, between late 1962 and 1966, which makes the type one of Solihull's rarer models.

Should you be interested in buying 695 PTA, William Trump can be contacted on 0404 822384. No time wasters or dealers please — on pain of seeing your name on a wanted poster in this column. For the purists, the vehicle no. is 28600116A and we believe from it's specification that it was originally destined for an export market.

110 Forward Control

Whilst on the subject of Series II Forward Controls, John Bowden of Gumtree Enterprises has sent photos of a South African military 110 FC. It is said that this vehicle was first used as a reserve ambulance then as a mobile field office, but personally, I think it was probably a communications vehicle.

Demobbed in 1990, it was shipped back to the UK, but the owner has now taken up a post in Malaya, and reluctantly had to part with it. The mileage of 5,029 is said to be genuine and for a vehicle in this type of role, it probably is.

For comparison, I've included a photo of a typical South African 110 Ambulance. although both vehicles are essentially similar, there are minor differences which suggest to me that Gumtree's vehicle is not an ambulance. In particular, the door in the left side, the sockets for a penthouse frame along this side and the ventilated side lockers (for batteries?) all suggest signals or command roles to me. Can anyone supply further information from South Africa please?

If you are interested in buying the South African Forward Control, Gumtree can be contacted on 0273 890259.

MILITARY SCENE

By Bob Morrison

△ Hard top FFR 110 in Germany

◁ Royal Air Force 110

▽ Soft top 110 FFR

△ *In addition to the usual radio antennae, this Army Corps communications 110 also carries a telescopic mast and tows a command post tent and equipment in the trailer.*

Well proven

FOLLOWING on from last month when we looked at the likely Solihull contender for the forthcoming Ministry of Defence Truck Utility Light contract, this month we look at the logical Land Rover candidate for TUM.

Truck Utility Medium (TUM) calls for a vehicle capable of carrying 1200kg at a sustained speed of 112 km/hr (70mph). Like TUL, the vehicle has to have a power to weight ratio of at least 20kw/tonne and a range of 450km at 80km/hr. Operating temperatures must be within the range of -31°C to +44°C and the power unit must run on Diesel and AVTUR.

Launched in 1983, and accepted for military service the following year, the 110 has now proved itself to be worthy of the Defender name tag. In the early days, there were some problems with the engines in one particular batch, but the performance of the 110 on Operation Granby (British code-name for the Gulf War) silenced most of Solihull's critics. As with the Challenger tank and Warrior personnel carrier, the Land Rover proved itself in combat, gaining praise even from General Sir Peter de la Billiere himself.

Sure, the military 110 didn't have air conditioning, quadraphonic sound systems and sun roofs, but it had the pulling power to drag itself through the unexpected quagmires of a rain lashed desert when fully laden and towing three times as much as the book laid down.

From the very beginning of British participation in the campaign to liberate Kuwait, the 110 was to the fore. It is thought that the first Land Rovers deployed probably belonged to an SAS unit working elsewhere in the Gulf region.

The full story of these Land Rovers will probably never be told, but I have it from the best sources that the Desert Patrol Vehicle 110s, nicknames Pinkies like their Series III predecessors, were later to the fore in the critical Scud-busting operations in Iraq.

Land Rovers also acted as mother ships for the small number of Light Strike Vehicles deployed. Incidentally, my sources also tell me that a number of Diesel 90s, nicknamed Dinkies, were specially converted for the SAS.

As a result of the Land Rover's sterling performance in the desert, US Special Forces now also use the 110 in quantity, but that is a story for another day.

However, it is wrong to attach too much importance to the "glamourous" SF roles of the Land Rover. By far the vast majority of service Land Rovers, beaver away almost un-noticed in the utility role.

With a payload close to 1500kg, seating for at least eight in the rear compartment and a road towing capacity of 4000kg. not to mention arguably the best off road performance of any vehicle in it's class, the 110 is the ideal General Service troop and cargo carrier.

However even the run-of-the-mill Land Rover can be expected to convert to specialist role. A large proportion of the 110 fleet are 24 volt and rigged for signals and communications duties, but they can quickly to stripped out for load-carrying or even medical duties if needed. Although 24 volt 110s are mainly hard tops, if operational requirements dictate, they can be easily stripped down to windscreen base level for airportability and airdrop.

At time of writing, around seventy nations use 110 Defender family vehicles in military and paramilitary roles. In UK service, 110s have proved themselves with the Royal Navy, the Royal Air Force and all Corps of the Army in all operating conditions.

EXERCISE ROADMASTER

By Tim Webster

Full of hope and expectation, The Brothers Websters in 'Higgy' at the start.

An early

THIS YEAR, LRO set out to win the Army's Roadmaster 4x4 event at all costs. So what went wrong? Tim Webster tells the tale.

IT HAD started, as if often does, with a phone call.

My brother, used to tough stuff in the British Army and obviously quite doubtful about my drunken claims last Christmas about driving Land Rovers up mountains, offered me a place on Roadmaster. 'Just your sort of thing,' he said in his way. 'Prove you've got what it takes, and all that. I'll navigate, got the maps and so on. You'll, um, have to, um, join the Army of course, but I did and I like it, so you will too., Forms in post.' I couldn't really say anything worthwhile before he rang off, and so promptly forgot all about it.

Three days later, the forms arrive — about thirty of them — with hand scribbled note from 'Bro. 'Can get you in on Special Enlistment because of prior TA service,' he wrote confidently.

The family, always very pro anything in green, then rallied round. 'You'll enjoy it' one parent said reassuringly. 'Show those blighters on that magazine you work for a thing or two,' said the other parent. 'You'll have to go along with it now, if only to save face' whispered my beloved, seeing the increasingly worried look on my face. 'S'pose so,' I agreed, reluctantly. 'Will I have to salute people, do you think?'. 'Yes, love, you probably will,' she sympathised.

And so five weeks, one lengthy and rather probing Army medical, a rapid hair cut, much form filling, and a rather embarrassing public strip and subsequent DPM kit dressage in the middle of a parade ground (the Army doesn't have changing rooms) later, Webster joined HM Armed Forces at its lowest level. 'Welcome to Salisbury Plain and Exercise Roadmaster,' boomed 'Bro. 'What did they say at the magazine when you told them you'd joined up to compete?' 'I said we'd win.' 'Oh hell,' was his reply…I think! I began to worry.

Exercise Roadmaster is probably the most popular Land Rover event in the British Army Motoring Association's (BAMA) sporting calendar. Run over an entire weekend, it's an on and off-road night navigation exercise and pretty demanding at that. Most of the driving takes place in the small hours when your energy is at its lowest ebb, and you cover upwards of 100 miles a night. Each evening, you're given a set of map segments and references (eight figure references for a 1:50,000 scale map, no less) and against these you plot a route 'corridor'.

But all that's the easy part. Along this route are manned Time Controls. Book in too early or late, and you're given time penalties. These add up to erode your maximum lateness allowance (just under an hour, cumulative) beyond which you're excluded from the event. Of the 120 crews that take part, the majority of retirements are in the first evening due to OTL or mishaps from mechanical failures, such is the critical pace of the event.

The vehicle park late on Saturday. Those not in uniform are members of Land Rover's team on hand to provide help and advice. ▷

Staff sergeants Hinds and Ellis of 4th Armoured Brigade brought this 110 V8 over from Germany to compete ◁

retirement

It was fairly clear from the two heavyweight crew briefings and subsequent map briefing that most of this was beyond my grasp that first evening. 'Bro had it spot on though, and marked up his maps as if he were creating a masterpiece. All I had to do was drive, which was, to my mind, the best part of the deal. Until the fog came down and I discovered 'Higgy'.

'Higgy', or 62 HG 77 to be precise, was our Land Rover for the event. Whereas everyone else seemed to have Ninetys or One-Tens (including a few V8's for the REME entries), we had the most dour, abused and dented Series 3 109" in NATO. The bloody thing didn't even sit square on its wheels, and the engine had a regular internal knocking that showed a bearing was out to lunch.

It was so embarrassing about Higgy that I shuffled around it, pretending it was nothing to do with me. Then, when no-one was looking, I stuck LRO stickers on the rear body sides (to keep 'Management' happy), only to find later that persons unknown had plastered OTTONS stickers over them. Since local Land Rover dealers OTTONS were supporting the event and every vehicle carried their stickers, it didn't really seem appropriate to make a fuss there and then. 'I'll show them,' I thought. 'Once on the winner's podium, I'll whip off the OTTONS stickers to reveal the LRO ones,' At that stage, the stress must have taken over, and I think I started talking out aloud. A passing Gurkha took a very wide berth.

22.30 hrs. Team Webster in Higgy, No. 92, headed for the start ramp to embark on Ex. Roadmaster. Unused to the clutch and low gearing of the 109", we hit the ramp a little quickly, bouncing the front wheels into the air. The start marshall stared at me. 'Cor, that's confidence for you. I didn't think you were going to stop. Mind you, No. 63 nearly fell off the ramp completely, so you're in good company.' I grinned weakly, 'Bro stared back and then pulled rank…and we were off.

The first 100 miles were a nightmare. The 109" twitched on greasy tarmac roads, we were harried and then overhauled by faster Ninetys, we got as far lost as 22kms off course, three TCs were found by accident (one while we were travelling at around 60mph), and off-road along Hampshire and Wiltshire byways, the fog obscured deep ditches until we found them…by hitting them at around 30mph. One we hit so hard that the jarring impact broke my highlift jack in the back of the Land Rover; the next one nearly broke my brother, who had his head through the narrow window aperture of his door at the time.

At the end of the first night, after hours of driving at 50mph or more into blanket fog, Team Webster was lying around 15th. Our navigation had been quite good, the driving indifferent, but we'd had the lucky break of trailing a Ninety crew that knew what they were doing. Most importantly Higgy had survived so far, his wheels no more crab tracked than they had been at the start and all this despite covering around 60 miles off-road at speeds at which I would have doubted my safety on-road!

EXERCISE ROADMASTER

EXERCISE ROADMASTER

As we slumped into sleeping bags in a continually lit gymnasium at the barracks — our sleeping and living area for the weekend — I remember thinking that army life was perhaps quite good after all.

That's until the dawn chorus from 200 waking soldiers started up. I got up when a Marine in a bunk not far from me joined in the chorus at around 8am with an unbridled 'pharp' to beat all those before. Day two had arrived.

Having fuelled Higgy's twin tanks and calculated our fuel consumption for the night before at something like 10-12mpg, we set off for the start point for the three daytime driving tests that intersperse the night-nav elements. On the first — a timed Slalom — we suffered a cruel blow as Higgy refused to answer accurately to the helm, and squashed more marked cones than anyone in the British Army could remember having been squashed before in one go.

We then embarked on Exercise Two — a Cross-Country Orienteering Exercise. This was pretty disastrous as well for Team Webster as it never dawned on us that you actually had to leave the vehicle to find the marker plates we so desperately needed to find. I think we ran over one or two in our haste, but they didn't count and nonetheless, we were sliding down the position board relentlessly.

Exercise Three was a thing called Regularity. You had to drive a marked out off-road course without stopping and in an exact time.

Onlookers were treated to Higgy hitting tank ditches at 50mph, coming down 1:1 slopes in neutral (the gearbox was playing up), pluming through mud traps and the constant wailing of the Websters trapped inside this seemingly possessed Army Land Rover.

At one point, our way was blocked by a bogged Ninety and the only course of action was to ram it and hope that sheer inertia would clear our way and prevent us stopping (which constitutes a fail). The plan worked, and Team Webster, No. 92, drove more sideways than ever through the Control Gate just four seconds over time, with Higgy only missing a number plate, front grille, headlight and bumperette.

I have to say here and now that the final Night-Nav stage that evening saw the early retirement of Higgy and Team Webster. Running a route 'corridor', mapped and controlled in the same way as the first evening, we simply went OTL because the fog prevented reasonably safe speedy progress. LRO's bid to be 'Roadmaster Champion' for 1992 had been thwarted by a bit of fog, a bitter pill to swallow because Team Webster had become a well-practiced dynamic pairing and Higgy was going better than ever, probably due to weight loss through shed bodywork.

The Tracked Vehicle Driver Training Area on Salisbury Plain, where the bulk of the Exercise Roadmaster Regularity Stage took place, consists of about nine square kilometres of criss-crossing tough tracks over undulating heathland. Although the tracks make full use of natural features to

The time-honoured tradition of entering water obstacles slowly, seems to be one not much favoured by members of the British Army Motoring Association, with predictable consequences for some, like Officer Cadet Chantale Mamane (right).

Photography: Bob Morrison

EXERCISE ROADMASTER

test driving ability, manmade trenches, pits and mounds are strategically placed to catch out even the most experienced driver. Surface texture varied from fine powder sand to coarse gravel, with peat and chalk thrown in for good measure.

As if this isn't enough, some of the pits and gullies are filled with standing muddy water which has the consistency of a bar of chocolate after two hours in a schoolboy's pocket.

Two water obstacles of particular interest were the Log Pool and the Mud Splash, the former being a natural ford with strategically placed log to catch the unwary and the latter a manmade pit.

Both were just deep enough to catch out the cautious driver and Bob Morrison spent the best part of the day wandering between one and the other as the field of 119 competitors tackled the Stage in relays.

As always on British Army Motoring Association events, entrants came from all three services and most Corps of the Army were represented.

Nearly one third of the teams were from the Territorial Army and seventeen teams travelled across from Germany for the event. BAMA events are one of the few occasions when a Colonel and Private can compete together as equals and it is not uncommon for the officer driving to be ordered about by the soldier navigating. Teamwork is definitely the order of the day.

Roadmaster is a 100% Land Rover event, and as always, the factory support team were on hand to offer advice and help to the contestants.

For the record, overall first place went to Gunner Braund and Sergeant Line of the Royal Artillery with 3 Fails and 87 Penalty points. Captains Paramore and Wilson of the Royal Army Ordnance Corps came second with 3 Fails and 106 Penalties.

Third place went to Corporal Hamilton and Sergeant Hampson of REME who travelled over from Germany to compete.

The joint RAF/Army team of Flight Lieutenant Fiona Burgess and Captain Bernie Stevens, who started in pole position, ran into trouble in the fog on the Saturday and finished outside the top ten.

On reflection, Roadmaster was a superb adventure, and one that shows just how tough Land Rovers are. Every vehicle in the event was pushed to its limits, and the number of retirements was nowhere near as high as you'd expect,.

By way of an epilogue, Higgy has probably joined the ranks of those Land Rovers waiting to be 'cast' through auction. 'Bro, suitably awed by my ability to drive a Land Rover through anything (even a stationary Ninety), has been notably quiet since. No doubt, he's cooking up another ruse. Me? Well, I'm committed to sixteen weekends a year military service, including two weeks annual training, my hair has got to be kept short and I'm financially responsible for my Army kit, half of which I left in the back of Higgy.

A high price for a weekend's off-roading? Well, no actually, because there's this BAMA event called Lowland Rover...!

• **Additional reporting: Bob Morrison.**

Airmobile gunners

By Bob Morrison

HEAVY LIFTING

WHEN I first started writing the military column in LRO, I had a shopping list of wanted photographs as long as my arm.

Over the years, the list has grown shorter as I've snapped subjects as diverse as Land Rovers leaving landing craft, Land Rovers under helicopters, Land Rovers on board ships, Land Rovers in aircraft, and Land Rovers on parachutes.

However, one shot that had always eluded me was a double lift of Solihull's babies under one helicopter.

During Exercise EAGLES EYE, 19 Field Regiment Royal Artillery put this right by laying on a double FC101 lift under an RAF Chinook from No 18 Squadron.

The Forward Control 101 Gun Tractor is now in the twilight years of it's military service and this could well have been the last double underslinging of the type. 19 Field Regiment, along with many other FC101 users, are soon to be re-equipped with the much delayed Reynolds Boughton RB44.

However, as the RB44 cannot be driven into a Chinook like a Land Rover, there is a good case for letting this airmobile artillery regiment keep their 101s until the new TUM (HD) enters service.

Without going into details, had the unit not been rehearsing tactics tailored to the introduction of the RB44, it is highly unlikely that two Land Rovers would have needed to have been carried externally, and I would not have got the shot.

Yorkshire based 19 Field Regiment of the Royal Artillery is the latest addition to 24 Airmobile Brigade. It's introduction now gives this rapid reaction formation it's own integral artillery support of 105mm light guns.

In addition to their gun tractors, the regiment is also equipped with a large number of Series III 109s, but these will soon be traded in for 110s.

Being an airmobile formation, everything is geared to keeping weight down so the Land Rover is the main source of transport when operational.

Having formerly been equipped with 155mm towed howitzers and the heavy trucks needed to tow and re-supply them, the lads of 19 Field are certainly finding their new role interesting.

SPECIAL OPERATIONS

Bob Morrison gets 'hands on' experience ▷

By Bob Morrison

Tough guys

HERALDED by a blaze of publicity in the Telegraph on the 20th of June, and pride of place on the British Aerospace stand at the EUROSATORY military exhibition in Paris two days later, the Land Rover Special Operations Vehicle is now out in the open. Solihull's Military Department actually gave LRO the opportunity to photograph the beastie, fully togged up, a couple of months ago, but as the press office embargoed my photos until after it's debut at Le Bourget on the 22nd of June, we were unable to do a full feature on it until this issue.

Although I made a couple of wrong diagnoses in my preview article on the SOV in the June issue, on the whole I seem to have got it pretty much right. The rectangular frames over the wings are for stowage boxes for mortar grenades and Claymores not Jerrycans, and the SOV seats six not five. The two front seats are conventional, but the rear acommodation consists only of seat squabs with lap belts. The demonstrator is not fitted with a strap seat for the main gunner, but it is highly likely that this would be fitted on production vehicles.

Weapons fit on the demonstration vehicle, which is actually the pre-production prototype, comprises mainly Royal Ordnance supplied equipment as, like Land Rover, they are part of the British Aerospace Group.

A copious aquantity of ammunition is stowed on the rear cargo bed, on top of the wheel boxes and in the front passenger footwell. The contents of the right wing stowage box are not known.

The SOV is based on a pretty much standard Defender 110 chassis with heavy duty suspension. One source tells me that a system for locking out the suspension to give even better weapons stability has been designed, but I am as yet unable to determine if this was fitted to the production batch.

The demonstration model sports a Southdown Underguard Protection System, as used by British Special Forces Land Rovers in the Gulf, and it is pretty safe to assume that this kit is fitted across the board. I believe that a Warn winch is standard on all SOVs and the in-service engine appears to be the 200Tdi.

Night convoy lights are fitted on the demonstrator, but it has been suggested by one contact that occulting beacons may be fitted on the production batch. The instrument panel is protected from the elements by a heavy duty canvas cover but the crew have no weather protection - if required this option is available, but the guys who will use this type of vehicle are probably not unduly worried by a bit of rain.

As far as can be determined at this stage, the SOV has been specifically designed as a readily airportable Rapid Intervention Vehicle. It does not have long range fuel tanks or masses of communications equipment as found on remote area or behind-the-lines patrol vehicles, but Solihull and the body manufacturer can offer the necessary fixtures for both if required by other clients.

If my evaluation is correct, two SOVs each carrying a six-man team, could be transported by Chinook, Hercules or similar aircraft. Time on-task will clearly be minimal as crew comfort has evidently been sacrificed for ease of exit and all-round firepower.

SPECIAL OPERATIONS

By Bob Morrison

Landing craft deliver Defender 110s to a south of England shore during a Royal Marine amphibious exercise.

GULF ROVERS

Marine land

MANY OF you will have noticed a recent glut of lightweights coming out on the civil market. The reason for this is that the last two major users of the type, namely 3 Commando Brigade and 5 Airborne Brigade, are now re-equipping with Defenders. Of course this does not mean that the airportable half-tonner has disappeared completely from military service, and no doubt odd examples will still soldier on until the next century, if past Land Rover form is anything to go by.

Recently I had the pleasure of photographing elements of 3 Commando Brigade with their new Land Rovers mounting an amphibious assault on a beach in southern England for the benefit of the visiting Secretary of State for Defence. Naturally, their new Defenders were to the fore.

The initial assault was carried out by Commandos in high speed Rigid Raider assault boats, followed by Royal Marines from 42 Commando in LCVP landing craft. If operational requirements dictate, two Defender 90s or a 90 and trailer can be brought ashore in an LCVP, but usually the first wave carries only troops to secure the landing.

The first Defenders, mainly winterised 110s, came ashore from larger LCU landing craft once the beach was secured. Each of the Royal Navy assault ships, HMS Fearless and HMS Intrepid, have large vehicle decks which access to their rear loading dock. Defenders reverse onto the 10 knot LCUs, each assault ship carries four LCUs, in the shelter of the dock and are then ferried ashore in a matter of minutes.

Like it's predecessors, the Defender can be rigged for deep wading, but, as this landing took place on a shingle beach, the vehicles rolled ashore in road going condition. Had the beach been gently sloping sand, with shifting sand bars, it is likely that waterproofing would have been necessary as a precaution. However, as prepared vehicles have to be de-rigged within the first few road miles to avoid mechanical damage, whenever possible amphibious landings are made bon steeply sloping beaches. This type of beach also decreases the risk of landing craft grounding and becoming sitting targets.

As the first Defenders go ashore on LCUs, self propelled pontoons known as Mexeflotes, are loaded with heavy plant and logistic vehicles from the fleet of Sir Bedi-

◁ *Airlift for a 101FC gun tractor*

▽ *101 ambulance comes ashore aboard a Mexeflote*

ing

vere class landing ships which are essentially roll-on roll-off vessels. One Tonne ambulances are usually brought ashore this way, while the One Tonne gun tractors of 29 Commando Regiment RA are underslung from the Commando Sea Kings which ply back and forth from the Aircraft Carriers, the assault ships, and the Royal Fleet Auxiliaries. The Defender can also be readily underslung, either on chains or in a net, but it is more usual for helicopter lifts to be reserved for guns, ammunition and gun tractors.

Although the Brigade's One Tonne FC101s should by now have been traded in for new vehicles, there is as yet no suitable in-service alternative, and it looks likely that the Royal Marines will end up keeping their gun tractors until TUM(HD) enters service in 1991 and their FC101 ambulances until the proposed All Wheel Drive Ambulance is available in mid-1995. By this time these vehicles will be around twenty years old and still running on petrol when the rest of the vehicle fleet is running on diesel.

As rapid response formations such as the Commandos, the Paras, ACE Mobile Force and 24 Airmobile Brigade all like the versatility of the FC101, particularly as a gun tractor, wouldn't it make more sense to refurbish and re-engine the low mileage examples now coming onto the civilian market in increasing numbers. A 200Tdi FC101 would be a formidable prime mover and could be in service in months instead of the years required to trial and introduce a new vehicle with all it's logistic back-up.

As for the new Defender – the lads all seem to like it,. Comfort wise, it certainly beats their old lightweights, and those who had been to Norway with the new vehicles couldn't praise the insulation and heating system enough. At the end of the exercise, they were all due for a spot of well earned leave, just as soon as their kit was back at base.

I used the same road back to Devon as the Commandos, but unlike years gone by, my journey was not a nightmare crawl behind convoys of overloaded, underpowered vehicles on the A35. For once the Brigade was able to roll along at the legal maximum.

UN FORCES

Nigerian 110 joins UN peacekeeping force in Croatia

UK MAMS 110 at Sarajevo Airport

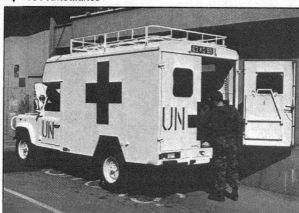

130 Ambulance

While bullets

WELL, either the bureaucrats listen to Paddy Ashdown, or they read LRO. As you'll see from the photos, last month's issue had hardly hit the news stands when the fleet of 127 ambulances was dispatched to Croatia to replace the battle weary 101s.

This batch of 127 ambulances, nowadays redesignated 130 by Solihull, are from the 63KG** Series and date from 1987/8. One of this Series was photographed by myself in the Central Vehicle Depot for a 1988 feature in LRO, and it is probable that this entire batch is from reserve stocks. It appears that they will replace the 101 fleet on a one-for-one basis, but with the official UN decision to deploy further British troops imminent as I write, I for one am not betting that all the 101s will be returned to the UK for some time.

Designed primarily as Airfield Crash Rescue Ambulances to work alongside the six wheel Range Rover TACR 2, this fleet is powered by the V8 and has superb performance on and off road. Unlike the 101 FC and the old 109 4-stretcher ambulances, road holding at speed is second to none and casualty comfort on rough roads is better than many conventional ambulances due to the coil suspension of the Defender family.

A common sight on all RAF and RN airfields, the 127 can carry either three stretcher cases or six to eight sitting patients or a combination load of two stretchers and four seating. In addition, there is a fold down seat for the attendant, which can be used by a casualty if the medic sits up front. This particular specification did not require a walk-through facility, but the casualty compartment is open to the cab almost to full width.

Internally, the ambulance has plentiful locker and stretcher stowage space beneath the seats and an overhead rack for blankets and the like, plus secure cabinets for medical supplies. The area above the cab in used mainly to stow the crew's kit. The large roof rack is primarily for the stowage of camouflage nets, but casualties' personal kit can be carried here if the ambulance has a full complement.

As far as I can determine from the photos, the only non-standard fitment on this batch is a radio communications kit. When Lt Col Lois Lodge, the Commanding Officer of 24 Airmobile Field Ambulance, was readying the original fleet of ambulances in the UK, she insisted that all her medics must be able to have radio contact with

130 and 101 ambulances side by side △

The BBC's armoured Land Rover 110 ▷

fly

their base and supporting troops. Believe it or not, military ambulances seldom carry even basic communications equipment in the field.

In addition to the medics, engineers and signallers already mentioned in past issues of LRO, many other units have Land Rover teams working on Operation Hanwood. Typical of this are the team from the Mobile Air Movement Squadron (UKMAMS) who unload the relief planes at Sarajevo Airport, when they get through. Often overlooked by the media, the Movers are right at the front line unloading aircraft in the sights of the sniper with mortar and howitzer fire an ever present danger.

Other nationalities are also using Land Rovers in United Nations colours in the former Yugoslavia, with the Kenyans even fielding locally produced 110s. In addition EEC monitors and UNHCR workers are also fielding Solihull's products, with the lead vehicle of the relief convoy into Gorazde being an armoured UN Range Rover.

All British military vehicles in Croatia and Bosnia at present are soft skin, but the BBC are using at least one, possibly two, fully armoured 110s. When Martin Bell got fragged in Sarajevo, he was outside his Glover Webb APV at the time, and radio reports of Kate Adie's tow-bruising encounter with a bullet suggested that it had bounced off her armoured Land Rover. A French TV team is running around in a Glover Webb Hornet and it is believed that other armoured Land Rovers ae in use by the media.

Why am I not out in Sarajevo you may ask. I've got no armoured Land Rover and a big yellow stripe running down my back.

This is a war where the Press are regarded as a better target than the opposition as hundreds of journalists with little military experience have found out to their cost.

The photos accompanying this article were kindly loaned by my good friend Laurie Manton who spent 10 days in the area for *Soldier* Magazine. In addition to being Assistant Editor of *Soldier*, Laurie is also one of the Territorial Army pool of Public Information Officers (TAPIO) and a keen amateur photographer. He recently won an Award of Excellence in the National Editing for Industry Awards for his Gulf War coverage.

Bob Morrison.

UN FORCES

By Bob Morrison

24 Airmobile Brigade 110 hard-top rigged for and awaiting a Chinook lift ▷

△ *Signals personnel shelter behind the soft-top which will be carried internally as the Chinook comes in to land. (Note the partially collapsed tilt)*

Soft-top 110 is loaded while the hard-top waits in front of the ▽ *helicopter*

Air mobile sig

AIRLIFTING

BRITAIN'S 24 Airmobile Brigade, as regular readers of this column will be aware, is primarily a helicopter borne formation tasked as an anti-armour reserve force for rapid deployment with NATO forces in Europe. Under NATO restructuring and the British Army Options for Change programme, this lightly equipped but heavily armed brigade will form the bulk of Britain's contribution to the Multinational Northern Airmobile Division (MNAD).

In the August issue, we took a brief look at the FC 101 Gun Tractors of 19 Field Regiment RA who provide the Brigade's integral artillery support. As a result, several readers have asked for more on the subject of underslinging and airmobility, so this month I'm featuring 210 Signals Squadron who provide the Brigade's communications.

In layman's terms, 210 SIGS provide the Brigadier and his staff with means of communication with a) the troops under his command, b) other formations working alongside and c) Divisional Headquarters. These communications means can be either radio, telephone or fax. The Squadron is also responsible for inhouse communications at Brigade HQ which allow, say, the Artillery Cell to speak to the Air Cell via secure land lines.

Usually the headquarters, known as Brigade Main, will be sited in a farm complex or similar collection of buildings where vehicles can be tucked away out of sight and a degree of protection from both the elements and enemy reconnaissance is provided. Alternatively a large thick wood will be used, but as Brigade Main receives visits from a lot of vehicular traffic which tends to churn up the access tracks, this location can be more liable to detection. As the battle progresses, Brigade Main must move with the action and the whole caravan is uprooted for maybe a fifty mile road march.

At no time during the move can the Brigadier afford to be out of touch with his superiors or subordinates, so a compact Fly Forward Headquarters (FFHQ) moves to the new location first to set up a communications centre. As helicopter lift availability is at a premium, 210 SIGS has shoehorned this entire set-up into four Land Rovers, which can be lifted by two Chinooks of the RAF Support Helicopter Force or, when working with the MNAD, two CH-53G Stallions of the German Army. Each helicopter carries personnel and one soft top 110 internally, with a hard top 110 & trailer combination underslung.

When Brigade Main is in operation, the FFHQ Land Rovers are parked up under cover a short distance away. They are fully rigged for helicopter lift and the crew wait with the vehicles on standby. Usually when

△ *The Chinook gently hovvers into position to collect the underslung hard-top and trailer combination*

▽ *The unit is whisked away to a new location*

nals

the Brigadier orders a move, the lads can expect about 90 minutes notice, but in an emergency they could be well on their way within 15 minutes.

The rear tilt frame of each soft top 110 is partially collapsed and tied down with paracord, which allows the vehicle to be parked right at the back of the Chinook, freeing maximum space up front for the soldiers. The hard top and trailer combination is permanently fitted with six chain slings, two front and rear on the bumper slinging points, and two on the rear spring hangers of the trailer.

Each pair of slings is of a set length to allow the combination to hang horizontal when underslung. Loading of the Land Rovers and trailers is critical to achieve the correct centre of gravity and balance.

Within 90 minutes of the order to move being given, the four Rovers are tucked into a woodline, overlooking the predetermined pick up point in a field, a couple of kilometres away from Brigade Main. As the distinctive wokka-wokka sound of the Chinooks is heard in the distance, the team break cover and line up the 110s in pairs roughly fifty metres apart and slightly staggered with the soft top about 50 metres directly behind it's hard top partner.

The Chinooks touch down briefly in front of the soft tops, and all but two men and the driver board at the double as the helicopter loadmaster drops the vehicle ramps. At the given signal, the soft top 110 drives straight up the ramp and the loadmaster secures it to the floor with chains - (before landing, he has stowed the rearmost troop seats to give adequate clearance for the Rover).

Less than a minute after the Rover's wheels stop turning, the Chinook is airborne. It hovers gently forward five metres above the ground as a soldier earths the hook with a pole then his mate slips the lifting ring into the Chinook's central load hook. The helicopter takes up the slack then whisks the team off into the distance.

Maybe thirty minutes later, the four 110s are 50 miles away at the new HQ location, and 15 minutes later they are camouflaged, up and running. The Brigadier can now transit by Command Helicopter to his new HQ as the rest of his staff, vehicles and equipment pack up and follow on by road.

Of the four Land Rovers at the new location, one is a Comms Vehicle (CV1) rigged with Ptarmigan, a second Comms Vehicle (CV2) is rigged with radios, the third (AMPAN) carries a Ptarmigan Access Node and the fourth (ARTY) carries the artillery communications equipment. Each vehicle weighs in close to the 3600kg legally permissible GVW, when fully rigged. Penthouses, stores and camouflage nets etc are carried in the trailers.

By Bob Morrison

Typical Rapier detachment Land Rover line up of two 130s and one 110 ▽

RAF Regiment

AT THE height of the Battle of Britain, Winston Churchill decreed that every man in RAF uniform should be armed and would be expected to fight and die in defence of his airfield. In the early days, cooks, clerks and ground crew all manned the guns, but by early 1942 the Air Force had formed it's own integral army, the RAF regiment.

Today, fifty years on, the RAF Regiment is still primarily tasked with airfield defence and covers both RAF Stations and field deployments. During the Gulf War, about two thirds of the Regiment's men and equipment were deployed to Saudi Arabia and the other front line countries to guard operating bases from surprise terrorist or Special Forces attack.

The bottom photograph on page 63 of the February 91 issue shows a typical Series III RAF Regiment patrol vehicle deployed at Dhahran air base to protect our Tornado detachment – at the time I could not disclose their location on security grounds. Other photos, of a 51 Squadron Series III can be found on page 30 of British Land Rovers in the Gulf.

On Operation Desert Sabre, the British ground offensive, RAF Regiment Gunners equipped with machine gun armed Land Rovers deployed into Kuwait. Although I have seen a few 110s, most photographs of RAF Rovers on Sabre show Series III 109 FFRs with tops and doors removed. All were fitted with at least one rear mounted GPMG, but a number also had forward firing machine guns on either a pintle mount or on the dashboard. Windscreens were generally fitted, but they were folded flat and covered with either hessian or cam nets.

See page 39 of the May 91 issue for a photo of one of the Series III Rovers during the big push.

The Regiment has two distinctive types of Squadron to counter attacks from either land or air. Earlier this year I visited 51 Squadron exercising in the Catterick area and 15 Squadron deployed on an airfield nearby, to get some insight into their quite separate requirements for Land Rovers.

At present, the land defence roled of Field Squadrons deploy with Scorpion light tanks, Spartan armoured personnel carriers and Land Rovers. Under the defence cutbacks in reply to the demise of the Warsaw Pact, their armour will soon be mothballed, and they will be re-equipped as primarily Land Rover borne formations.

Gradually, the Series IIIs that they use for reconnaissance and patrol duties are being replaced by Defender 110s which are re-equipped with machine gun mounts and other fitments stripped from the older

△ Field Squadron Defender 110 FFR

▽ Field Squadron 110 recce and patrol vehicle

models. Hard top Defender 110 FFRs are also now entering service with the Regiment for command and communications duties.

Each recce or patrol vehicle carries four personnel armed with a light support weapon and three SA-80 individual weapons. For heavy support, the Rovers are equipped with both front firing and rear pintle mounted general purpose machine guns.

A normal FFR kit is carried in the rear along with copious quantities of ammunition, fuel and water. A man-pack radio for use in dismounted action is carried between the front seats. The Land Rover retains it's roll cage but is otherwise stripped down to windscreen base level for reduced silhouette, and it's shape is broken up with hessian and cam nets.

The Rapier Squadrons, who provide air defence cover also deploy with Land Rovers. A typical Rapier Detachment is equipped with two special bodied Defender 130 models to tow the launcher and fire control units, transport personnel and carry missile reloads. Both of the Rapier Tractors are fitted with winches to facilitate easy positioning of the launcher and radar.

More than 200 of the 130 Rapier Tractors are in British service and at least one ally also uses this model for airfield defence on NATO's eastern flank. Incidentally, the RAF Regiment's 130s were purchased after a long and hard competitive trials programme to find the best available vehicle for this particular role.

The Rapier Detachment's third vehicle is a hard top Defender 110 FFR used for communications duties and to tow a trailer containing their personal kit.

Whilst most Middle and Far East users of the Rapier system deploy 110s in pairs to cover tractor and re-supply roles, the dedicated 130 allows the RAF and USAF to deploy an optically controlled launcher plus missile reload with only one vehicle when operational requirements dictate. However, if necessary, the detachment's 110 can double as a Rapier Tractor.

In addition to protecting conventional airfields, the RAF Regiment also regularly deploy into the field in support of Harriers and helicopters working from Forward Operating Bases. These FOBs are usually set on the edges of woods or forests where the aircraft can be hidden under camouflage nets away from prying eyes and aerial surveillance. This is when the Land Rovers really come into their own, both for positioning or re-supplying the missile batteries and undertaking recce or defensive patrols off the beaten track.

CROOK BROTHERS

SPECIALIST IN EX-MINISTRY OF DEFENCE VEHICLES FOR OVER 30 YEARS
BLACKBURN OLD ROAD, HOGHTON, PRESTON, LANCS PR5 0RX
(Only 5 miles from M6 junction 28 or 29 or from M61 junction 8). *See directions below*

FINE SELECTION OF LIGHTWEIGHT LAND ROVERS
Mostly 12 volt and no VAT. Few 24 volt ffr, canvas top or fitted plain hard-top, or with side windows, rear seats, rear door etc. Nato Green, matt or glossy, Light Stone or Black/Green Cam.

110 DIESEL HARD-TOPS
New MOD release, very clean, rear seat,s (no VAT), on Michelin XCL tyres, in gloss Nato Green, ex-RAF.

109" FIRE CRASH RESCUE TENDER T.A.C.R.I.
With pump, water tank, hoses, foam nozzles, 900x16 Trak Grip tyres.

SWB SERIES III DIESEL
Hard-top, windows, rear door, also 4-cyl. Petrol canvas top, or as photo.

LWB SERIES III LAND ROVERS, DIESEL & 4-cyl. PETROL, RHD & LHD IN STOCK
As canvas top, plain hard-top, or with side windows, rear seats, rear door (no VAT). Nato Green, Black/Green Cam. or Light Stone.

LIGHTWEIGHT LAND ROVER
Refurbished in Desert Storm colours, 24 volt ffr.

LAND ROVER 101 1-tonne FORWARD CONTROLS
Choice available, as received from M.O.D. Unprepared or checked and prepared to high standard. Ready to drive away on age related number plates. In stock, RHD & LHD, all 12 volt, some fitted with winch.

LAND ROVER AMBULANCE SER. III, 1981
In exceptionally clean condition, only showing 8,000 miles, fitted for four stretchers, ex-RAF, exellent off-road ambulance, or ideal for camper/overland expedition, military spec., twin fuel tanks etc.

RANGE ROVER AMBULANCE
Very clean throughout, direct M.O.D. (ex-Royal Navy), Lomas body, fitted for 2 stretchers, full rear door, exellent camper/expedition vehicle etc. Tested/licensed fitted number plates, ready to drive away.

ALSO IN STOCK
Bedford Mk 4x4 Diesel Trucks, Bedford TK Turbo Diesel 7.5 Truck, jerricans, ammo boxes, spares etc. Any vehicle prepared with 12 months Test 6 months Road Tax, age related no. plates, export enquiries welcome. Shippng arranged to any port.
SELECTION OF VARIOUS TYPES OF M.O.D. SANKEY TRAILERS, CARGO BODY, MOBILE RADIOGRAPH, DARK ROOM WITH HIGH TOP, SMALL OFFICE etc.

As you leave motorway follow brown/white historic monument signs for "Hoghton Tower", we are just past entrance on the right next to "The Boars Head" on the A675.

OPEN MONDAY-FRIDAY 8am-12pm & 1pm-5.30pm • SATURDAY 9am-4pm • SUNDAY by appointment

Tel: 0254 852660 & 0254 852457 • Fax: 0254 853334

MILITARY VEHICLE SPECIALISTS

WE'VE MOVED!

SUPPLIERS TO THE FOREIGN & COMMONWEALTH OFFICE

KEITH GOTT

GREENWOOD FARM, OLD ODIHAM ROAD, ALTON, HANTS GU34 4BW
(Take road out of Alton by side of Crown Hotel on High Street — then 1½ miles on left)

SPECIALISTS IN EX-MOD, CIVILIAN AND EX-AUTHORITY LAND ROVERS

"All Vehicles can be purchased in 'AS IS' condition/or fully prepared with a years MoT, full service and registration"

ALL MODELS OF LAND ROVER & RANGE ROVER SERVICED AND REPAIRED

LAND ROVERS EXPORTED TO OVER 45 COUNTRIES

SERIES III SWB, fully refurbished. Price Guide £4000-£6500

RANGE ROVERS Price Guide £4000-£14,000

110 DIESEL MOD Fully refurbished £6000-£8000

LT/WT SERIES III Military colours & trim £2500-£3500 Fully refurbished £4000-£6000

101 FORWARD CONTROL £3000-£6500+VAT

90 COUNTY STATION WAGON WITH 2.8 ISUZU TURBO DIESEL £12,500

110 DIESELS & PETROLS Petrol from £3250+VAT (Civvy) Diesels from £4750 No VAT (Mod)

Farmers and Foreign Office, campers and caravanners, exports and expeditions — we deal with them all. Basic running/driving models to deluxe vehicles built to your specification. **THE PRICE DEPENDS ON YOU.** Some are undersealed from new, have twin fuel tanks, oil coolers and H/D suspension amongst their many extras.

LET US ADVISE YOU — THAT COSTS NOTHING
EXPORT ENQUIRIES WELCOMED

Tel: (0420) 544330, Fax: (0420) 544331

SWB/LWB SERIES III Ex-MOD from £2500

MILITARY & CIVILIAN

EX-MOD & CIVILIAN USED-VEHICLE SALES
LAND ROVER, RANGE ROVER
FERRET, SARACEN, ABBOT
DDR G-WAGEN, TRAILERS
PLANT ETC.

WITTER ◆

COMPETITIVELY PRICED QUALITY PARTS & ACCESSORIES FOR ALL MODELS

◆ WARN WINCHES

MAIL ORDER
PERSONAL CALLERS
LOCAL DELIVERIES

DIXON BATE ◆ NATO HITCHES!

SEND FOR LATEST CATALOGUE!

SELECTION OF VEHICLES AVAILABLE:
LAND ROVERS - ALL TAX, TESTED & PLATED
S3, LWB, 24v FFR, From - £2,200+VAT
S3, LWB, 12v, From - £2,000
S3, Light Weights, From - £2,500
Ex-MOD Trailers, From - £450+VAT

FERRET SC's, 4WD, 6 Cyl. Petrol, Camo, From - £1,500+VAT
HUMBER PIG APC, Fully Restored, Beautiful Example - £2,500
CHIEFTAIN TANKS, £P.O.A.
ABBOT Self Prop Guns, £P.O.A.

Unit C, Ringstones Ind. Est., Bridgemont, Whaley Bridge, Stockport, Cheshire, SK12 7PD

TEL/FAX: 0663 734042

THE LIGHTWEIGHT CENTRE LTD.

"Simply the Best"

We are the only company in the UK which specialises in the Lightweight and we use our 16 years experience to select only the best of those released from the MOD.

Prices range from £2000 to £4500 inc. Full MOT

Tel: (0823) 666505 or 0860 926572
for full details

EX-MINISTRY LAND ROVERS

refurbished and ready for work.
TO INCLUDE:
● LWB ● LIGHTWEIGHTS
● AMBULANCES
● SWB AND 24 VOLT EQUIPMENT

SPARES, SERVICING, WELDING & REPAIRS

SAFARI ENGINEERING
EVERSLEY, North Hampshire
Tel: (0734) 732732
Mobile: 0836 384505

BROOKLYN ENGINEERING
HURSLEY RD., CHANDLERS FORD, NR. SOUTHAMPTON, HANTS. SO5 1JH
EX MINISTRY OF DEFENCE LAND ROVER'S — BEDFORD'S — TRAILERS
HOME MARKET EXPORT SPECIALIST FOR OVER 25 YEARS

LAND ROVER — DISCOVERYS — RANGE ROVER S/3 — 90 — 110 — COUNTYS PETROL/DIESEL. We try to keep a good selection of vehicles to suit most pockets from the S/3 for towing the horse box to Discoverys for private use 110 Countys for private use or that trip overseas. We would also be interested in P/X on your L/R. All vehicles checked in our own workshops.

EX MOD L/ROVER (Direct from MOD). Land Rovers RHD-LHD. S/3 — Cutwings — 101s — Ambulances from as is condition to fully prepared to your spec. including MOT reg. no. (year related). If you want a L/Rover for war games or overseas trip home market and export we can help with all transport and shipping.

EX-MOD SANKEY ¾ TON TRAILERS. 2-wheeled, over-run brake, hand brake, landing legs, as is condition to rebuilt water/diesel bowzer 300 gallon (new tank) fitted on trailer chassis, prices from £250-£400, bowzers from £600-£700+VAT.

DISCOVERY TDI. 1991 (H), white, alloy wheels, 7 seater, electric pack, 27,000 miles, taxed, very clean .... £16,000 (no VAT)

DEFENDER TDI. 1990 (H), green, full county 12 seater spec., 36,000 miles, fully serviced, clean, PAS £12,000+VAT

90 COUNTY Turbo Diesel, white, full spec., MoT, very clean, but high mileage, one owner, MOT'd .. £8,400 (no VAT)

101 AMBULANCE LHD. 26,000 kl (direct from MOD), mint condition, MOT'd, serviced etc .... £7,000+VAT

101 SOFT TOP RHD. V8 with winch Michelin radial tyres, mint condition. (This vehicle has been on Jersey), Nato green, must be seen, MoT etc .... £6,000 (no VAT)

LIGHTWEIGHTS. Ex MOD 2 RHD 2 LHD, nice clean genuine vehicles from ...... **from £2,400 to £3,250 (no VAT)**

EX MOD LAND ROVERS. We try to keep a good selection of 101"-109"-88". L/weights please phone/fax for details.

109" LHD STATION WAGON. Petrol 12 volt, 1980 (W), green, 10 seater, needs tidying, MoT'd £1750 (no VAT)

LAND ROVER — RANGE ROVER SPARES
New and Used — Genuine/Pattern Parts EX-MOD & CIVILIAN

We carry all parts new and used for Ex MOD & Civilian S/3 — 90 — 110 — Range Rover Discovery.
V8 ENGINE (NEW) FACTORY UNIT RTC 2382N 8.25 com. (new price over £2000) £998.75
PETROL ENGINE 2.6 (6 cylinder). Complete. MOD recon £1057.50
WEBER CARB KITS (for 2¼ Petrol) £76.37
CARPET KITS (black), 4-cylinder S/2, S/3 .... £41.12
110 LEVELLING UNIT (Gen.) NRC 7050 .... £146.87
110 FRONT SHOCKERS (Gen.) RTC 4637 .... £14.68
CUTWING TOP BULKHEAD, New (Ex-MOD). Series 2 type £88.12
Series 3 type £146.88
101 WINCH. Complete kit, s/hand £881.25
TYRES. Goodyear/Avon, all size new used
ACCESSORIES & EQUIPMENT — Bull bars, FW Hubs, Winches, roof Racks, Jerri Cans, Overdrives, Chassis parts, Workshop manuals. All above prices include VAT at 17½%. Please phone or call into our stores they will be only too pleased to help with new/used parts. Post/Packing etc., extra (at cost price).
EXPORT/OVERSEAS — Enquiries welcome large or small, please use our 24hr Fax No. All models L/Rover, R/Rover, Discovery, 12volt/24volt spares (ex MOD). **FRIENDLY SERVICE YOU CAN BE SURE OF**

BEDFORD DIESEL ENGINE. 330 (6 cylinder). Direct from MOD, fully recon. complete ready to install in M/Type (inline injector pump) including clutch .... **£1,250+VAT**

BEDFORD 4 x 4. M/Type RHD/LHD. Diesel engine drop side/chassis cabs as is condition or to your spec., one off LHD with Atlas Crane, 4,000 kl only

MOTOR BIKE. Armstrong 500cc, 1985. Ex MOD 9,000 kl complete bike, fully checked, resprayed sand £950
(For sale on behalf of owner)

AMMO BOXES (Metal/Wood)
PENETRATING OIL WD40 type, gallon cans
JERRI CANS New/Used
WATER CANS Plastic New/Used
TOW ROPES 10' long, eye each end (Manilla)
SHOVELS Long/short handle
PICK AXES Long/short handle
CAMOUFLAGE NETS 24' x 24' as new, cot. type
JUNGLE KNIVES with holders.
Nato green paint Gal. Cans (genuine) Ex. MOD

0703-252281 0703 269990 (Fax 24 hrs.) More details, photos etc. on request

---

THE WILTSHIRE LAND ROVER CO. LTD.
Snarlton Lane, Melksham, Wilts.

That Wiltshire Land Rovers is around here somwhere!!

New & Used Spares always available

Tel. 0225 700975
0831 277048 (23½ hr.)

EX-MINISTRY OF DEFENCE

LAND ROVER 110" L.H.D. 2.5 L PETROL. QTY. 40 AVAILABLE EX-STOCK.

Qty. 20. L.H.D. SERIES 3, 109", full canvas, low miles. Excellent condition.

Also good selection of Good Quality Lightweight, 101, 88", 109", 90" & 110" Military vehicles. Petrol and Diesel.

SPECIAL OFFERS
Snow Chains .... £117.50 per pair inc. VAT
Lightweight Series 2A
 Bulkhead Vent Panel ...... £117.50 inc. VAT
Rover Diffs. 3.54 Ratio .... £235.00 inc. VAT

UK carriage paid

L. JACKSON & COMPANY
Tel. (0302) 770485/491.
Fax: (0302) 770050

BROOKLANDS ROAD TEST SERIES

Abarth Gold Portfolio 1950-1971
AC Ace & Aceca 1953-1983
Alfa Romeo Alfasud 1972-1984
Alfa Romeo Alfetta Coupés GT. GTV. GTV6 1974-1987
Alfa Romeo Giulia Berlinas 1962-1976
Alfa Romeo Giulia Coupés Gold Portfolio 1963-1976
Alfa Romeo Giulia Coupés 1963-1976
Alfa Romeo Giulietta Gold Portfolio 1954-1965
Alfa Romeo Spider Gold Portfolio 1966-1991
Alfa Romeo Spider 1966-1991
Allard Gold Portfolio 1937-1959
Alvis Gold Portfolio 1919-1967
American Motors Muscle Cars 1966-1970
Armstrong Siddeley Gold Portfolio 1945-1960
Aston Martin Gold Portfolio 1972-1985
Austin Seven 1922-1982
Austin A30 & A35 1951-1962
Austin Healey 100 & 100/6 Gold Portfolio 1952-1959
Austin Healey 3000 Gold Portfolio 1959-1967
Austin Healey Sprite 1958-1971
BMW Six Cyl. Coupés 1969-1975
BMW 1600 Collection No. 1 1966-1981
BMW 2002 1968-1976
BMW 316, 318, 320 (4 cyl.) Gold Portfolio 1975-1990
BMW 320, 323, 325 (6 cyl.) Gold Portfolio 1977-1990
BMW 5 Series Gold Portfolio 1981-1987
Bristol Cars Gold Portfolio 1946-1992
Buick Automobiles 1947-1960
Buick Muscle Cars 1965-1970
Buick Riviera 1963-1978
Cadillac Automobiles 1949-1959
Cadillac Automobiles 1960-1969
Cadillac Eldorado 1967-1978
Chevrolet Camaro Z28 & SS 1966-1973
Chevrolet Camaro & Z28 1973-1981
High Performance Camaros 1982-1988
Camaro Muscle Portfolio 1967-1973
Chevrolet 1955-1957
Chevrolet Corvair 1959-1969
Chevrolet Impala & SS 1958-1971
Chevrolet Muscle Cars 1966-1971
Chevelle & SS 1964-1972
Chevelle & SS Muscle Portfolio 1964-1972
Chevy Blazer 1969-1981
Chevy El Camino & SS 1959-1987
Chevy II Nova & SS 1962-1973
Chevrolet Corvette Gold Portfolio 1953-1962
Chevrolet Corvette Gold Portfolio 1968-1977
Chevrolet Corvette Sting Ray Gold Portfolio 1963-1967
High Performance Corvettes 1983-1989
Chrysler 300 Gold Portfolio 1955-1970
Chrysler Valiant 1960-1962
Citroen Traction Avant Gold Portfolio 1934-1957
Citroen 2CV 1948-1988
Citroen DS & ID 1955-1975
Citroen SM 1970-1975
Cobas & Replica 1962-1983
Shelby Cobra Gold Portfolio 1962-1969
Cobras & Cobra Replicas Gold Portfolio 1962-1989
Daimler SP250 Sports & V-8 250 Saloon Gold Portfolio 1959-1969
Datsun Roadsters 1962-1971
Datsun 240Z 1970-1973
Datsun 280Z & ZX 1975-1983
De Tomaso Collection No. 1 1962-1981
Dodge Charger 1966-1974
Dodge Muscle Cars 1967-1970
Excalibur Collection No. 1 1952-1981
Facel Vega 1954-1964
Ferrari Cars 1946-1956
Ferrari Dino 1965-1974
Ferrari Dino 308 1974-1979
Ferrari 308 & Mondial 1980-1984
Ferrari Collection No. 1 1960-1970
Motor & T&CC Ferrari 1966-1976
Motor & T&CC Ferrari 1976-1984
Fiat-Bertone X1/9 1973-1988
Fiat Pininfarina 124 & 2000 Spider 1968-1985
Ford Consul, Zephyr, Zodiac Mk.I & II 1950-1962
Ford Zephyr Zodiac Executive Mk.III & Mk.IV 1962-1971
Ford Cortina 1600E & GT 1967-1970
High Performance Capris Gold Portfolio 1969-1987
High Performance Fiestas 1979-1991
High Performance Escorts MkI 1968-1974
High Performance Escorts MkII 1975-1980
High Performance Escorts 1980-1985
High Performance Escorts 1985-1990
High Performance Sierras & Merkurs Gold Portfolio 1983-1990
Ford Automobiles 1949-1959
Ford Fairlane 1955-1970
Ford Ranchero 1957-1959
Thunderbird 1955-1957
Thunderbird 1958-1963
Thunderbird 1964-1976
Ford Falcon 1960-1970
Ford GT40 Gold Portfolio 1964-1987
Ford Bronco 1966-1977
Ford Bronco 1978-1988
Holden 1948-1962
Honda CRX 1983-1987
Hudson & Railton 1936-1940
Jaguar and SS Gold Portfolio 1931-1951
Jaguar XK120, XK140, XK150 Gold Portfolio 1948-1960
Jaguar Mk.VII, VIII, IX, X, 420 Gold Portfolio 1950-1970
Jaguar Mk.2 1959-1969
Jaguar Cars 1961-1964
Jaguar E-Type Gold Portfolio 1961-1971
Jaguar E-Type 1966-1971
Jaguar E-Type V-12 1971-1975
Jaguar XJ12, XJ5.3, V12 Gold Portfolio 1972-1990
Jaguar XJ6 Series II 1973-1979
Jaguar XJ6 Series III 1979-1986
Jaguar XJS Gold Portfolio 1975-1990
Jeep CJ5 & CJ6 1960-1976
Jeep CJ5 & CJ7 1976-1986
Jensen Cars 1946-1967
Jensen Cars 1967-1979
Jensen Interceptor Gold Portfolio 1966-1986
Jensen Healey 1972-1976
Lagonda Gold Portfolio 1919-1964
Lamborghini Cars 1964-1970
Lamborghini Countach & Urraco 1974-1980
Lamborghini Countach & Jalpa 1980-1985
Lancia Fulvia Gold Portfolio 1963-1976
Lancia Stratos 1972-1985
Land Rover Series I 1948-1958
Land Rover Series II & IIa 1958-1971
Land Rover Series III 1971-1985
Land Rover 90 & 110 1983-1989
Lincoln Gold Portfolio 1949-1960
Lincoln Continental 1961-1969
Lincoln Continental 1969-1976
Lotus Elite 1957-1964
Lotus Elite & Eclat 1974-1982
Lotus Elan Gold Portfolio 1962-1974

Lotus Elan Collection No. 2 1963-1972
Lotus Cortina Gold Portfolio 1963-1970
Lotus Europa Gold Portfolio 1966-1975
Lotus Turbo Esprit 1980-1986
Motor & T&CC on Lotus 1979-1983
Marcos Cars 1960-1988
Maserati 1965-1970
Maserati 1970-1975
Mazda RX-7 Collection No. 1 1978-1981
Mercedes Benz Cars 1949-1954
Mercedes Benz Cars 1954-1957
Mercedes Benz Cars 1957-1961
Mercedes 190 & 300 SL 1954-1963
Mercedes 230/250/280SL 1963-1971
Mercedes Benz SLs & SLCs Gold Portfolio 1971-1989
Mercedes S and 600 1965-1972
Mercedes S Class 1972-1979
Mercury Muscle Cars 1966-1971
Metropolitan 1954-1962
MG Gold Portfolio 1929-1939
MG TC 1945-1949
MG TD 1949-1953
MG TF 1953-1955
MG Cars 1959-1962
MG Midget 1961-1980
MGA & Twin Cam Gold Portfolio 1955-1962
MG Midget 1961-1980
MGB Roadsters 1962-1980
MGB MGC & V8 Gold Portfolio 1962-1980
MGB GT 1965-1980
Mini Cooper Gold Portfolio 1961-1971
Mini Muscle Cars 1961-1979
Mini Moke 1964-1989
Mopar Muscle Cars 1964-1967
Morgan Three-Wheeler Gold Portfolio 1910-1952
Morgan Plus 4 & Four 4 Gold Portfolio 1936-1967
Morgan Cars 1960-1970
Morgan Cars Gold Portfolio 1968-1989
Morris Minor Collection No. 1 1948-1980
Shelby Mustang Muscle 1965-1970
Shelby Mustang Muscle Portfolio 1965-1970
Mustang Muscle Cars 1967-1971
High Performance Mustangs 1982-1988
Oldsmobile Automobiles 1955-1963
Oldsmobile Cutlass & 4-4-2 1964-1972
Oldsmobile Muscle Cars 1964-1971
Oldsmobile Toronado 1966-1978
Opel GT 1968-1973
Packard Gold Portfolio 1946-1958
Pantera Gold Portfolio 1970-1989
Panther Gold Portfolio 1972-1990
Plymouth Barracuda 1964-1974
Plymouth Muscle Cars 1966-1971
Pontiac Tempest & GTO 1961-1965
Pontiac Muscle Cars 1966-1972
Pontiac Firebird & Trans-Am 1973-1981
High Performance Firebirds 1982-1988
Pontiac Fiero 1984-1988
Porsche 356 1952-1965
Porsche Cars in the 60's
Porsche Cars 1960-1964
Porsche Cars 1964-1968
Porsche Cars 1968-1972
Porsche Cars 1972-1975
Porsche 911 1965-1969
Porsche 911 1970-1972
Porsche 911 1973-1977
Porsche 911 Carrera 1973-1977
Porsche 911 Turbo 1975-1984
Porsche 911 SC 1978-1983
Porsche 914 Collection No. 1 1969-1983
Porsche 914 Gold Portfolio 1969-1976
Porsche 924 Gold Portfolio 1975-1988
Porsche 928 1977-1989
Porsche 944 1981-1985
Range Rover Gold Portfolio 1970-1992
Reliant Scimitar 1964-1986
Riley 1.5 & 2.5 Litre Gold Portfolio 1945-1955
Rolls Royce Silver Cloud 1955-1965
Rolls Royce Silver Cloud & Bentley 'S' Series Gold Portfolio 1955-1965
Rolls Royce Silver Shadow 1965-1981
Rover P4 1949-1959
Rover P4 1955-1964
Rover 3 & 3.5 Litre Gold Portfolio 1958-1973
Rover 2000 & 2200 1963-1977
Rover 3500 1968-1977
Rover 3500 & Vitesse 1976-1986
Saab Sonett Collection No.1 1966-1974
Saab Turbo 1976-1983
Studebaker Gold Portfolio 1947-1966
Studebaker Hawks & Larks 1956-63
Avanti 1962-1990
Sunbeam Tiger & Alpine Gold Portfolio 1959-1967
Toyota MR2 1984-1988
Toyota Land Cruiser 1956-1984
Triumph TR2 & TR3 1952-1960
Triumph TR4, TR5, TR250 1961-1968
Triumph TR6 Gold Portfolio 1969-1976
Triumph TR7 & TR8 1975-1982
Triumph Herald 1959-1971
Triumph Vitesse 1962-1971
Triumph Spitfire Gold Portfolio 1962-1980
Triumph 2000, 2.5, 2500 1963-1977
Triumph GT6 1966-1974
Triumph Stag 1970-1980
TVR Gold Portfolio 1959-1990
VW Beetle Gold Portfolio 1935-1967
VW Beetle Gold Portfolio 1968-1991
VW Kubelwagen 1940-1975
VW Karmann Ghia 1955-1982
VW Bus, Camper, Van 1954-1967
VW Bus, Camper, Van 1968-1979
VW Bus, Camper, Van 1979-1989
VW Beetle Collection No.1 1970-1982
VW Scirocco 1974-1981
VW Golf GTI 1976-1986
Volvo PV444 & PV544 1945-1965
Volvo Amazon-120 Gold Portfolio 1956-1970
Volvo 1800 Gold Portfolio 1960-1973

BROOKLANDS ROAD & TRACK SERIES

Road & Track on Alfa Romeo 1949-1963
Road & Track on Alfa Romeo 1964-1970
Road & Track on Alfa Romeo 1971-1976
Road & Track on Alfa Romeo 1977-1989
Road & Track on Aston Martin 1962-1990
Road & Track on Auburn Cord and Duesenburg 1952-1984
Road & Track on Audi & Auto Union 1952-1980
Road & Track on Audi & Auto Union 1980-1986
Road & Track on Austin Healey 1953-1970
Road & Track on BMW Cars 1966-1974
Road & Track on BMW Cars 1975-1978
Road & Track on BMW Cars 1979-1983
Road & Track on Cobra, Shelby

& Ford GT40 1962-1992
Road & Track on Corvette 1953-1967
Road & Track on Corvette 1968-1982
Road & Track on Corvette 1983-1986
Road & Track on Corvette 1986-1990
Road & Track on Datsun Z 1970-1983
Road & Track on Ferrari 1950-1968
Road & Track on Ferrari 1968-1974
Road & Track on Ferrari 1975-1981
Road & Track on Ferrari 1981-1984
Road & Track on Ferrari 1984-1988
Road & Track on Fiat Sports Cars 1968-1987
Road & Track on Jaguar 1950-1960
Road & Track on Jaguar 1961-1968
Road & Track on Jaguar 1968-1974
Road & Track on Jaguar 1974-1982
Road & Track on Jaguar 1983-1989
Road & Track on Lamborghini 1964-1985
Road & Track on Lotus 1972-1981
Road & Track on Maserati 1952-1974
Road & Track on Maserati 1975-1983
Road & Track on Mazda RX7 1978-1986
Road & Track on Mazda RX7 & MX5 Miata 1986-1991
Road & Track on Mercedes 1952-1962
Road & Track on Mercedes 1963-1970
Road & Track on Mercedes 1971-1979
Road & Track on Mercedes 1980-1987
Road & Track on MG Sports Cars 1949-1961
Road & Track on MG Sports Cars 1962-1980
Road & Track on Mustang 1964-1977
Road & Track on Nissan 300-ZX & Turbo 1984-1989
Road & Track on Peugeot 1955-1986
Road & Track on Pontiac 1960-1983
Road & Track on Porsche 1951-1967
Road & Track on Porsche 1968-1971
Road & Track on Porsche 1972-1975
Road & Track on Porsche 1975-1978
Road & Track on Porsche 1979-1982
Road & Track on Porsche 1982-1985
Road & Track on Porsche 1985-1988
Road & Track on Rolls Royce & Bentley 1950-1965
Road & Track on Rolls Royce & Bentley 1966-1984
Road & Track on Saab 1972-1992
Road & Track on Toyota Sports & GT Cars 1966-1984
Road & Track on Triumph Sports Cars 1953-1967
Road & Track on Triumph Sports Cars 1967-1974
Road & Track on Triumph Sports Cars 1974-1982
Road & Track on Volkswagen 1951-1968
Road & Track on Volkswagen 1968-1978
Road & Track on Volkswagen 1978-1985
Road & Track on Volvo 1957-1974
Road & Track on Volvo 1975-1985
Road & Track - Henry Manney at Large and Abroad

BROOKLANDS CAR AND DRIVER SERIES

Car and Driver BMW 1955-1977
Car and Driver BMW 1977-1985
Car and Driver on Cobra, Shelby & Ford GT40 1963-1984
Car and Driver on Corvette 1956-1967
Car and Driver on Corvette 1968-1977
Car and Driver on Corvette 1978-1982
Car and Driver on Corvette 1983-1988
Car and Driver on Datsun Z 1600 & 2000 1966-1984
Car and Driver on Ferrari 1955-1962
Car and Driver on Ferrari 1963-1975
Car and Driver on Ferrari 1976-1983
Car and Driver on Mopar 1956-1967
Car and Driver on Mopar 1968-1975
Car and Driver on Mustang 1964-1972
Car and Driver on Pontiac 1961-1975
Car and Driver on Porsche 1955-1962
Car and Driver on Porsche 1963-1970
Car and Driver on Porsche 1970-1976
Car and Driver on Porsche 1977-1981
Car and Driver on Porsche 1982-1986
Car and Driver on Saab 1956-1985
Car and Driver on Volvo 1955-1986

BROOKLANDS PRACTICAL CLASSICS SERIES

PC on Austin A40 Restoration
PC on Land Rover Restoration
PC on Metalworking in Restoration
PC on Midget/Sprite Restoration
PC on Mini Cooper Restoration
PC on MGB Restoration
PC on Morris Minor Restoration
PC on Sunbeam Rapier Restoration
PC on Triumph Herald/Vitesse
PC on Spitfire Restoration
PC on Beetle Restoration
PC on 1930s Car Restoration

BROOKLANDS HOT ROD 'MUSCLECAR & HI-PO ENGINES' SERIES

Chevy 265 & 283
Chevy 302 & 327
Chevy 348 & 409
Chevy 350 & 400
Chevy 396 & 427
Chevy 454 thru 512
Chrysler Hemi
Chrysler 273, 318, 340 & 360
Chrysler 361, 383, 400, 413, 426, 440
Ford 289, 302, Boss 302 & 351W
Ford 351C & Boss 351
Ford Big Block

BROOKLANDS RESTORATION SERIES

Auto Restoration Tips & Techniques
Basic Bodywork Tips & Techniques
Basic Painting Tips & Techniques
Camaro Restoration Tips & Techniques
Chevrolet High Performance Tips & Techniques
Chevy Engine Swapping Tips & Techniques
Chevy-GMC Pickup Repair
Chrysler Engine Swapping Tips & Techniques
Custom Painting Tips & Techniques
Engine Swapping Tips & Techniques
Ford Pickup Repair
How to Build a Street Rod
Land Rover Restoration Tips & Techniques
Mustang Restoration Tips & Techniques
Performance Tuning - Chevrolets of the '60's
Performance Tuning - Pontiacs of the '60's

BROOKLANDS MILITARY VEHICLES SERIES

Allied Military Vehicles No.1 1942-1945
Allied Military Vehicles No.2 1941-1945
Off Road Jeeps: Civ. & Mil. 1944-1971
US Military Vehicles 1941-1945
Complete WW2 Military Jeep Manual
US Army Military Vehicles WW2-TM9-2800